DIGITAL ASSET

VALUATION FRAMEWORK

by HashKey Capital

The Commercial Press

Executive Editor: Hockey Yeung
Cover Designer: Cathy Chu
Typesetter: Zhou Rong
Printing Advisor: Kenneth Lung

Digital Asset Valuation Framework

Author: HashKey Capital
Editors-in-Chief: Xiao Feng Aaron Low Deng Chao David Zou
Editor: HashKey Capital Research Team:
Henrique Centieiro Scarlett Xiao Jupiter Zheng
Publisher: The Commercial Press (H.K.) Ltd.
8/F, Eastern Central Plaza, 3 Yiu Hing Road, Shaukeiwan, Hong Kong
Distributor: SUP Publishing Logistics (HK) Limited
16/F, Tsuen Wan Industrial Centre, 220-248 Texaco Road, Tsuen Wan, N.T., Hong Kong
Printer: Elegance Printing and Book Binding Co. Ltd.
Block A, 4th Floor, Hoi Bun Industrial Building
6 Wing Yip Street, Kwun Tong, Kowloon, Hong Kong

First edition, First printing, April 2024

ISBN: 978 962 07 6734 0 (Hardcover)
978 962 07 6735 7 (Paperback)
Printed in Hong Kong, China

Table of Contents

Chapter 2 Public Chain/Layer 1 Valuation

Chapter 3 DAO Tokens and Utility Tokens

Chapter 4 A Framework for DAO Token Valuation

Chapter 5 STO Valuation

Chapter 6 NFT Valuation

Preface

From the initial birth of Bitcoin in 2008 to the present day, blockchain technology and the cryptocurrency market have been rapidly evolving, becoming a powerful force in the global financial realm. There is a strong interest and great attention towards this new type of asset known as cryptocurrency. The approval of 11 Bitcoin spot ETFs by SEC in early 2024 further signifies the formal shift of cryptocurrencies from the periphery to the mainstream. Investors can now trade Bitcoin like stocks, transforming it into an asset that traditional investors can comprehend.

Although cryptocurrencies have been in existence for several years, they are still an early-stage industry, both in terms of technological development and asset management. Cryptocurrencies differ from mainstream assets in that the value of most cryptocurrencies is derived from consensus rather than a standardised valuation framework. These consensus-based cryptocurrencies have created both wealth effects and some scepticism from the mainstream. Questions arise, such as why cryptocurrencies hold value, whether Bitcoin has any underlying support, what drives cryptocurrency prices, and what reasons there are to invest in cryptocurrencies. However, as the industry gradually matures, we are discovering that traditional financial valuation methods can be applied to the valuation of cryptocurrencies. This enables traditional institutional investors to understand the industry, attracting more capital and giving rise to more unicorns, similar to what happened in the internet and consumer industries.

At this momentous time when Bitcoin spot ETFs have been approved in the United States, and a new bull market is on the horizon, HashKey Capital has decided to write this book. HashKey Capital is a venture capital that has been deeply involved in blockchain investments for nearly a decade. We have invested in over 500 projects covering various sectors of the industry, with the aim of capturing value across the entire ecosystem. Throughout this journey, we have focused on ecosystem development, developer support, and project investments, growing alongside the industry. Our understanding has continuously improved as we leverage our insights and investment logic from the primary market, combining traditional financial experience with blockchain and cryptocurrency investment activities. This has led us to develop a framework for valuing cryptocurrency assets. We now share this valuation framework, hoping to provide reliable valuation methods for cryptocurrency investors.

As one size doesn't fit all, the digital asset valuation framework outlined in this book provides guidance to evaluate any asset within the framework.

First, we categorise the different types of crypto assets with the help of our token categorisation tree.

Second, we apply the token valuation matrix that allows us to identify a token valuation method that fits the crypto asset that one is trying to evaluate.

Third, we apply the token valuation method to the token being evaluated.

For clarity, this book aims to bring a crypto asset valuation and not an evaluation. People use the words valuation and evaluation interchangeably, but some important differences exist. Evaluation refers to assessing something; for example, evaluate if a product is good, evaluate a brand, evaluate how tasty the food is in that new Michelin restaurant or the potential of a certain investment return. On the other hand, valuation refers specifically to determining an asset's financial value. Valuation involves analysing data to determine the value of an asset. In this book, we focus on the second.

It is also important to say that valuation and prices reflect two different things. While valuation represents the intrinsic value of something, the price is the result of many other variables that often differs from the value: sentiment, market hype or depression, speculation, fear, greed, overinflated news, etc. We can affirm that the prices of most assets in the world have a non-negligible percentage of irrationalism, but crypto sometimes suffers from this disorder to a higher degree. Specifically, there are many factors that can influence cryptocurrency prices, including market demand, market sentiment, government policies, regulatory changes, technological innovations, industry partnerships, project progress, and so on.

Throughout this book, the reader will stumble upon valuation methods that might provide indicative price projections. Still, it's important to note that these price projections are based on valuation metrics that encompass a small number of variables. Consequently, when applying these metrics, one always needs to take into account that there are many other "ceteris paribus" variables.

Long-term successful investing in crypto doesn't require an unusual IQ, special business insights, or insider information. Instead, what's needed is a solid framework for making investment decisions based on value rather than speculation.

We sincerely hope that this book can provide you with assistance in your crypto asset investments, answer your questions, and offer support and guidance. As you read through this book, we hope you can deeply engage with its content and enjoy the journey of cryptocurrency investment. We wish you a pleasant reading experience!

About HashKey Capital

HashKey Capital is a global digital asset and blockchain leader helping institutions, founders and talents advance the blockchain industries.

As one of the largest crypto funds and the earliest institutional investor in Ethereum, HashKey Capital has over US$1 billion AUM since its inception, with over 500 investments in infrastructure, tools, and applications.

With our deep knowledge across the blockchain ecosystem, HashKey Capital has built a robust network connecting founders, investors, developers, and regulators.

Introduction

If you are reading this book, we are sure you have come across questions such as: Why does crypto have value? Is Bitcoin backed by anything? What moves crypto prices?

With this book, we hope to provide the crypto community, investors, and traditional finance with a valuation framework that allows them to find those answers.

We also hope the book provides crypto investors with additional valuation tools for their fundamental analysis to level up crypto investments with traditional securities in terms of analytical rigour and understanding.

In this book, we outline a framework that allows us to evaluate different crypto assets consistently. Successful crypto investing requires a solid framework for value-based decisions rather than speculation.

The efficient-market hypothesis would tell us that prices already include all the information about the asset into account and that the price reflects the exact value of the asset. Asset prices instantly change to reflect new information. Therefore, market prices are perfect, and it is impossible to outperform the market using information that the market already knows.

However, we could provide thousands of examples that go against it. The reality is that markets are inefficient, and we have thousands of examples in the crypto and stock markets.

Why did Apple's stock price decline by 30% in Feb. 2020 during the Covid crash? Did Apple sell 30% less products? Did Apple have 30% less earnings? Why did GAP stock crash 70% in 49 days? Did the biggest clothes retailer in the US suddenly sell 70% fewer clothes? The answer to all these questions is a resounding no. Markets are often irrational, and prices will not reflect value. As Warren Buffett said, "Mr. Market is a drunken psycho."

This is where the concept of value investing comes from. Value investing tries to identify securities that are underpriced due to market irrationality. These securities are trading below market value (i.e., with a margin of safety), and investors who buy them, hope that one day, the price will reflect the value of the stock. Quoting the greatest investor Warren Buffett of all time again: "Price is what you pay, value is what you get."

The same happens with crypto assets. The Bitcoin price crashed by 50% during the Covid crash, only to recover to the previous levels 55 days later and double its price in the consequent 5 months. Did these movements in price accurately represent the real value of the network, its activity, active wallets, transaction volume, and network usefulness? Did the prices reflect the value? The answer is no. The trick is to have the right tools to spot these opportunities, which we attempt to do in this book.

The market is like a chess tournament. All chess players have access to the same information and read the same chess books. However, there will always be a small number of chess players that will outperform and use the information available with more effectiveness to win the game.

This book is a chess rule book but for crypto asset valuation, and it will help the reader to use and interpret information already available to the market to his advantage. We have built a framework to bring clarity to the value of crypto assets.

Intrinsic value and market price:

There are a few axioms that are very important to any asset, including crypto assets, and although crypto assets are a relatively new asset class, they inherit similar core principles:

- A crypto asset is not just a ticker symbol backed by nothing and created out of thin air. In most cases, crypto assets represent an underlying network, utility, and an actual decentralised business.
- The market forever swings between bull sentiment and bear sentiment, between greed and fear, between unsustainable optimism and unjustified pessimism. Valuation methods are important to maintain a rational investment strategy.
- The future performance of an investment is a function of its present price compared to the valuation.
- It is possible to minimise risks by avoiding overpaying and disregarding overhyped projects by looking at their valuation metrics.

Even though this market has existed for a little longer than ten years, we try to give practical examples and case studies with statistical significance. Saying this, there are still many caveats and limitations as we can only access a few years of data, and the market is still evolving, with new regulations being implemented every year.

We hope that this book brings some clarity to the industry, and we would be happy to review it in the future as the market evolves and we gather more data that will help us understand the intrinsic value of crypto assets.

One final word: this book stands on the shoulders of giants. We do not intend to invent the wheel. We are simply refurbishing valuation metrics and methods that were built by the giants of traditional finance and equities analysis and adapting them to crypto assets.

Token Categories

There are different categories of crypto assets that require different treatment in terms of asset valuation methodology.

In this section, we lay down the different types of digital assets and a decision tree that can be applied to categorise the crypto assets and provide additional guidance to categorise these assets.

For each crypto asset, there are different asset valuation methodologies that can be used, which we outline in the table below. Remember that unlike securities, cryptocurrencies and tokens might have a wide range of purposes and utility and give investors different rights. Given the diversity of crypto assets in the market, there's no "one-size-fits-all" valuation method. The investor needs to look at the different valuation methods and see which best fits each specific case.

Token category tree:

Is the asset fungible? → No → **NFT**

🠋 Yes

Is the asset's main purpose paying transaction fees on the blockchain?
→ Yes → **Payment tokens/cryptocurrencies/L1/L2 and native crypto**

🠋 No

Is the token used for governance? → Yes → **DAO tokens**

🠋 No

Is the token used for fee payment, loyalty points, or a reward program? →
Yes → **Utility tokens**

🠋 No

Does the token represent security (stocks, bonds, etc.) → Yes → **STO**

Crypto asset valuation framework matrix

There are a number of different methods/models that aim to evaluate digital assets. The crypto asset valuation framework will help us find which methods/models best fit the corresponding digital asset type. After identifying the type of asset that you are trying to evaluate, find the respective models in the table below. Then, feel free to navigate to the respective section of the book. Some of the models are suitable for different types of cryptographic assets.

Framework matrix	
Crypto asset	Models analysed
Native crypto (e.g., Bitcoin)	• Value of replacing traditional financial institutions • Comparison with telecom networks • BMLC – Bitcoin's Moore Law Compensation • Bitcoin price equilibrium adjusted to M2 • Stock to flow model • P/S ratio • Bitcoin comparison with the gold ETF • Bitcoin cost method • Transaction value to transaction fee • Metcalfe's Law • NVT ratio • Market approach
Smart Contract L1s (e.g. Ethereum)	• Value of replacing traditional financial institutions • Comparison with telecom networks • Valuation and price equilibrium according to M2 • Stock to flow model • Transaction value to transaction fee • DCF - Discounted cash flow • Value of a perpetual bond • Metcalfe's Law • NVT ratio • P/S ratio • Market approach
Utility Tokens	• Discounted cash flow • Market approach • Comparable tokens • QTM – Quantity Theory of Money • Market capitalisation to TVL

Framework matrix	
Crypto asset	Models analysed
DAO Tokens	• Discounted cash flow • NCAVPT - Net Current Asset Value Per Token • Market approach
STO	• Market approach • Income-based approach and DCF • Cost approach
NFTs	• Market approach

Chapter 1

Valuation of Native Cryptocurrencies

In this section we look at first generation cryptocurrencies. Being the most widely used Bitcoin, there are other first-generation cryptocurrencies such as Litecoin, Dogecoin, and other Bitcoin forks.

The primary goal of first-generation cryptocurrencies is to allow for simple peer-to-peer transactions and to be used as a store of value. After the creation of Bitcoin and the validation of its usefulness, other first-generation blockchains were created by splitting up from the Bitcoin chain. These chain split-ups typically happen when a group of miners/nodes agree that a new chain with new rules should be created. In this case, the two groups of miners will split up. One will continue to support the blockchain with no changes, and the other will support the blockchain with the changes. This kind of split is called a fork.

An early example of this is the Bitcoin–Litecoin fork. In late 2011, a developer called Charlie Lee proposed the implementation of a new chain that would split (fork) from the Bitcoin blockchain. While Bitcoin uses the SHA-256 hashing algorithm, 10-minute blocks with a size of 1MB, he proposed that a new blockchain called Litecoin would adopt a different hashing algorithm called Scrypt, a 2.5-minute blocks instead of 10-minute and 256kb size blocks instead of 1 MB.

Litecoin, as well as many other Bitcoin forks like Bitcoin Cash and other forks of forks like Dogecoin, live until today and can be included in the first generation of cryptocurrencies or native cryptocurrencies.

First-generation native cryptocurrencies main use cases are narrowed down to sending value across space (from person A to person B) and preserving value across time. Bitcoin's limited supply ensures that there's no debasement over time, thus being an asset that is seen by many as a good store of value.

As we will see later in this section, Bitcoin derives its value not only from its primary use cases but also from the demand that comes from the facts that Bitcoin is the most secure computer network ever created, decentralised with no single point of failure and no possible censorship and the most scalable, with its layer 2 allowing for potentially millions of transactions that leverage the Bitcoin network as the settlement layer.

In this section, we will talk about different approaches to the valuation of native cryptocurrencies.

The table below gives an overview of the valuation methods outlined in this chapter, their respective goal, and their applicability. Although in this chapter we use Bitcoin as the target asset, most of these valuation methods can be applied to other assets, as detailed in the table.

Valuation method	Goal	Applicability
Value of a trusted decentralised network replacing centralised institution	Comparing the utility of the crypto asset and its decentralised network with traditional institutions and estimate its value	Large chains and ecosystems such as Bitcoin and Ethereum
Bitcoin comparison with telecommunication networks	Comparing the crypto asset with the value of other large networks	Large chains and ecosystems that benefit from network effects
Bitcoin price equilibrium adjusted to M2	Calculating the price equilibrium of the crypto asset by leveraging the correlation with the M2 money supply	Large chains and ecosystems such as Bitcoin and Ethereum
Bitcoin Stock to Flow Model	A method to value scarce assets according to their existing stock and flow	Crypto assets with limited supply
Bitcoin P/S ratios	A method that correlates the asset price and sales (in the form of fees/rewards) and infer its relative value	Crypto assets that generate revenue, from L1 and L2 chains, to utility tokens and DeFi tokens
Predicting Bitcoin Price from the Perspective of Bitcoin ETFs	Measuring the impact of an ETF demand on the price of the crypto asset	Any crypto asset with a spot ETF approved in a major market
Cost of production method	Measuring the cost of production of a crypto asset and comparing it to its price	Proof of work crypto assets
Transaction value to transaction fee	Measuring the efficiency and cost-effectiveness of a network according to its utility	Bitcoin and large chains

Each of these approaches will deliver different results, but the goal is to provide the investors with methodologies that allow for better cryptocurrency analysis and comparable, as well as understanding the intrinsic value of digital assets.

Even though digital assets, including Bitcoin, have only a few years of price history to backtest, our valuation methods bring investors a systematised and standardised approach to their valuation.

Time will tell if the valuations provided in this book are accurate, but our approach is the first effort in the industry to systematise and standardise the valuation of cryptocurrencies.

1.1 The value of a trusted decentralised network replacing centralised financial institutions

Networks such as Bitcoin can do the same job as a financial institution when it comes to transferring and storing value across the world. Consequently, it is logical to value Bitcoin as an asset that, just like banks, captures value from facilitating transactions and being a store of value.

Although in this section we will focus on Bitcoin, other chains have use cases that mimic a wide range of financial products and non-financial products. The emergence of smart contracts allows us to fully automate the operations of a business without having to rely on new infrastructure. With infrastructure, we mean servers, computers, office space, corporate infrastructure, organisational infrastructure, and employees.

Smart contract compatible chains like Ethereum enabled the existence of DeFi. In DeFi, there is a multitude of examples of use cases - from lending, decentralised exchanges, derivatives, insurance, etc. - however, the most straightforward example for comparison with the financial industry is still Bitcoin. Although we do not forget that some Layer 1 blockchains can also store any kind of data, which greatly expands their potential use cases, they might have too many use cases to use a simple comparable valuation.

Bitcoin, on the other hand, has straightforward use cases, thus allowing us to compare it to the centralised equivalents in the financial industry more easily.

Bitcoin has no employees, offices, suppliers, or any sort of fixed costs tied to the transactions that it facilitates. Saying this, it would be fair to attribute it a value according to the network's usefulness. If Bitcoin can have some of the functions of a bank, what would be its fair value based on those functions?

We call this the comparable value of a trusted decentralised network replacing centralised financial institutions.

Bitcoin brings the level of trust that banks used to bring to the financial industry, but in this case, in a decentralised manner.

An important thing to consider is that Bitcoin is simultaneously a network, an internet protocol, and a cryptocurrency that can be used to send and store value, as well as reward the network nodes/ miners that keep the network secure. This is quite different from any other asset class in the world.

We call "unity" the fact that Bitcoin is the same asset that is used to transact and incentivise the stakeholders involved in the process. This unity property brings digital assets a recursive effect and a self-adjusting effect, where the value of the asset will directly impact all its stakeholders, including the miners who keep the network secure. Consequently, when the value of the asset rises, the incentive to secure the network as a miner also rises. This is why an increase in the mining participants (usually measured as network hashrate) is seen as an increase in the value of the network.

To understand unity even better, we can do a quick thought experiment. Let's use the Visa network as an example.

Visa shares do not have the same "unity" property because we cannot use Visa shares to transact on the Visa network, and Visa cannot pay for the maintenance of its servers with Visa shares and will not pay salaries to its employees with shares.

If hypothetically, Visa shares are used not only to represent the ownership of the company and for governance but also to be used as a currency and payment itself on the Visa network and to incentivise the Visa stakeholders (suppliers, users, employees, etc.), Visa shares would have a much bigger use case, better velocity and Visa shares would be worth much more.

Another big advantage of unity is incentive alignment across all the stakeholders. While a centralised business might have many misaligned parties, there's a full alignment across all the Bitcoin stakeholders. For instance, a supplier that sells or maintains Visa computers and servers will always have an incentive to overcharge for its services whenever possible, while some Visa employees will work less than what they are paid for.

On the other hand, Bitcoin's unity brings total incentive alignment, which is 100% regulated by the market prices and the immutable Bitcoin code.

As a result, we propose a direct comparison between native cryptocurrencies (in this case, Bitcoin) and the traditional financial market (in this case, the banking industry).

To make this comparison and understand the size of the Bitcoin network relative to the size of the banking industry, we can use a ratio that derives from their market capitalisation.

BTC/Banking ratio = BTC market capitalisation/
banking market capitalisation worldwide

This will give us a ratio that allows us to compare the BTC capitalisation (and consequently price) with the capitalisation of the banking system.

A higher BTC market capitalisation compared to the banking market capitalisation (in this case, expressed by the BTC/Banking ratio) translates into higher usage of the Bitcoin network, and it should demand a higher Bitcoin price.

If the banking market capitalisation grows and the BTC/Banking ratio remains the same or grows, then we should see an increase in the BTC price.

Conversely, if the BTC/Banking ratio declines, we should see a decline in the Bitcoin price.

The table below shows historical data for the BTC market capitalisation, the banking sector market capitalisation, the BTC/Banking ratio, and the BTC price.

Date	BTC Market capitalisation (trillion USD)	Banking Market capitalisation (trillion USD)	Ratio BTC/ Banking	BTC price
2016	0.016	6.8	0.24%	$864
2017	0.32	7.9	4.05%	$16,900
2018	0.056	6.5	0.86%	$3,200
2019	0.0132	7.6	0.17%	$7,300
2020	0.46	6.1	7.54%	$24,700
2021	0.9	7.8	11.54%	$47,600
2022	0.32	7.7	4.16%	$16,800
2023	0.88	7.5	11.73%	$42,488

Source: Statista

Considering our main assumption, which is the fact that Bitcoin can perform part of the functions performed by the banking system, we have outlined the BTC/Baking ratio using the Bitcoin market capitalisation and world's banking system market capitalisation.

We could stop our analysis here and simply use the BTC/Banking ratio as a metric to understand if the asset is underpriced or overpriced in relation to the banking sector, but we can also try to make some future price predictions.

One could assume that if, at any point in the future, the BTC/banking ratio increases, but the Bitcoin price doesn't increase, the Bitcoin price is undervalued. The same vice-versa.

Conclusion

Bitcoin and other cryptocurrencies present a transformative approach to traditional financial systems, offering a decentralised and efficient platform for transactions and value storage.

The comparable value of Bitcoin as a decentralised network replacing centralised financial institutions provides a unique perspective in assessing its value. The unity property of Bitcoin, which is its simultaneous functioning as a network, protocol, and cryptocurrency, sets it apart from any other asset class.

The proposed BTC/Banking ratio, calculated by comparing Bitcoin's market capitalisation to the global banking market capitalisation, offers a method to evaluate Bitcoin's growth and potential. Given Bitcoin's inherent qualities and its potential to perform functions currently fulfilled by traditional banks, this ratio and corresponding price projections serve as compelling indicators for Bitcoin's future valuation.

1.2 Bitcoin comparison with telecommunication networks

The comparison method can be used for Bitcoin valuation, as well as for utility tokens, as we will detail later in the utility token section. Although in the previous example, we compared Bitcoin to the financial industry, and in this example, we will compare Bitcoin to the telecommunication industry, other cryptocurrencies/tokens might be compared to other industries. For example, Filecoin is a decentralised storage protocol. At the time of writing this book, Filecoin stores an amount of data equivalent to 10% of all AWS storage (AWS – Amazon Web Services is the biggest cloud provider in the market). Saying this, we could compare Filecoin's market capitalisation to the AWS market capitalisation or even compare it with the total cloud storage market capitalisation.

Bitcoin can be compared to gold - as it is very often compared by the Bitcoin community, as well as prominent economists such as Saifedean Ammous, the author of *The Bitcoin Standard*. We also drew some gold comparisons in this section of the book.

We can not only compare Bitcoin to gold as an asset, but we can also eventually compare the Bitcoin network to the global telecom network. Note that Bitcoin is a brand new economy, and it's 100% internet-based and decentralised, and, just like telecom networks, Bitcoin is all about storing data (which coincidentally, in this case, represents monetary value).

To compare Bitcoin to the telecom industry, we would have to gather the global telecom market cap and make a comparison to Bitcoin. However, it is important to note that unlike Bitcoin, telecom stocks that encompass the global telecom market cap don't have the unity element that Bitcoin has, i.e., a share in the Bitcoin network is the same unit that is used to transact.

The top telecom companies have large market capitalisations (2023):

China Mobile ($182 billion), AT&T ($140 billion), Verizon ($176 billion), and T-Mobile ($186 billion) are just some of the examples. The aggregated market capitalisation of the 150 largest telecom companies in the world is $2.6 trillion. The market capitalisation of the telecom sector has tripled over the last 20 years.

Considering that at the time of writing this book, the Bitcoin market capitalisation is around $800 billion, it is possible to draw a comparison between the value of the Bitcoin network and the value of the telecommunications network, which sits at $2.6 trillion.

Saying this, for Bitcoin to reach the same size as the telecommunications network, the market capitalisation would have to grow 3.25 fold.

Considering the rigid Bitcoin supply, the impact on Bitcoin price would likely be more than 3.25 fold. To calculate the Bitcoin price compared to the telecommunications market, we would recommend using the price equilibrium method of supply and demand (later discussed in this book).

Bitcoin and telecommunications networks have obvious differences, and comparing them has strong limitations. However, it is still possible to draw historical comparisons between them, as well, from the network utility perspective.

1.3 Hallmarks of value and equilibrium of a decentralised network of assets

As in any other asset class that exists in a free market, there's a multitude of reasons why prices move, and the price of any asset is the result of thousands of people buying and selling, and they do so for many different reasons. However, we lay down the main key demand drivers for Decentralised Networks of Assets.

On the demand side, Bitcoin price increases with the number of network participants and how much they value the network in terms of:

- Risk aversion and censorship aversion
- Demand for decentralised networks
- The value of the network services with network effects
- Quality of the products transacted
- Future network participation
- Demand for censorship resistance
- Demand for spanning across time and space
- Digital alternative to precious metals

- Demand for trustlessness with the elimination of counterparty risk, transaction time/cost, verification costs, and legal costs
- Demand for transparency and eliminating information asymmetries.
- Demand as e-cash proof of concept
- Demand for a cheap payment network
- Demand for a programmable shared database of value
- Demand for an uncorrelated financial asset

According to Nick Carter, General Partner at Castle Island Ventures and prominent Bitcoiner, Bitcoin narratives and evolving uses can also evolve over time.

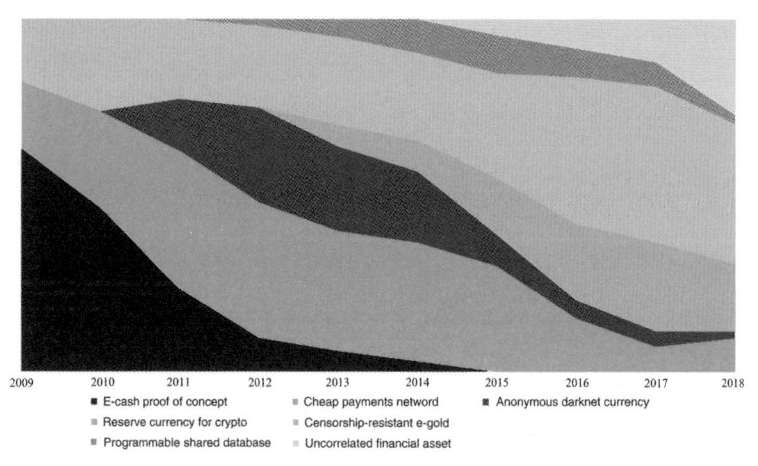

Source: Visions of Bitcoin blog post by Nick Carter on Medium.

All these factors are part of demand D. D can be a function of all the demand factors D= f(A, B, C, D, F…).

The equilibrium price is where Demand and Supply meet. D = S

In other words:

D = [Risk aversion and censorship aversion] +
[Demand for decentralised networks] +
[Risk aversion and censorship aversion] +
[Demand for decentralised network] +
[The value of the network services with network effects] +
[Quality of the products transacted] +
[Future network participation] +
[Demand for censorship resistance] +
[Demand for spanning across time and space] +
[Digital alternative to precious metals] +
[Demand for trustlessness with the
elimination of counterpart+ y risk] +
[Demand for transparency and
eliminating information asymmetries] +
[Demand for an uncorrelated financial asset] + [...]

Knowing the demand and supply of Bitcoin, we can easily calculate the price using the price equilibrium equation (later detailed in the price equilibrium section in this book).

$$P = qS/qD$$

While the demand comes from all these factors, the supply is limited by the Bitcoin code. Unlike in other industries where new market competitors will increase the supply of the product (for example, new car manufacturers will add new supply in the market), in the case of Bitcoin, new miners in the market will not increase the Bitcoin supply. New miners will increase only the hashrate, which can be a measure of network security.

The Bitcoin supply is highly inelastic, making the price more sensitive to demand variations, volatile, and less risky in the long term.

1.4 Bitcoin's Moore Law Compensation

In this section, we will break down the correlation between the amount of computing power miners contribute to the network – also called hashrate – and the Bitcoin price. We will also include Moore's law in the calculation.

Historically, there's a positive correlation between Bitcoin's hashrate and Bitcoin price. However, it's important to highlight that correlation doesn't necessarily always mean causation. Typically, new Bitcoin miners are drawn to the network when the Bitcoin price increases (or it's expected to increase) because then, the mining rewards will be higher. However, we can also see an increase in the hashrate as a long-term bet in Bitcoin's price.

The correlation can easily be identified with a linear regression on a logarithmic scale, which is very indicative of the tight correlation between these two variables. However, we also need to be aware of multicollinearity occurring in the model. Multicollinearity, in this case, is good as it allows us to access highly correlated metrics – price and hashrate – giving us a framework for Bitcoin valuation and price prediction.

One of the arguments as to why the Bitcoin value is directly linked to the hashrate is that a higher hashrate means more security for the Bitcoin network, making it more attractive as a reserve asset.

Although small variations of the Bitcoin hashrate (let's say, up to 10%) don't have a meaningful relative increment/reduction of the Bitcoin network security, it is important to understand whether the net computing power (and consequently security) is increasing or decreasing over time.

To better understand how effectively the Bitcoin hashrate evolved, we need to look at net hashrate gains compared to how much microprocessing technology evolved in that same period of time. This is where we need to add Moore's law into the calculation.

Moore's law states that computer chip capacity doubles every two years. Suppose this law continues to be verified to measure how Bitcoin security evolved. In that case, we need to measure how the hashrate behaved during that same period by compensating for Moore's law. Did the hashrate grow faster or slower than Moore's law?

Should we use the Moore's law? A lot of people criticised Moore's lay by saying that it cannot always be observed (for example, Intel took five years to advance from 14 nm chips to 10 nm). However, Moore's Law isn't necessarily about the size of a chip but rather about the speed that it offers. There are many variables for speed/capacity other than chip size and transistor count. For this reason, Moore's 1995 paper doesn't limit Moore's law to transistor count. There are other chip improvements that can increment their capacity. Saying this, we will assume in this section of the book that Moore's Law will continue in the future.

Bitcoin's Moore Law Compensation – BMLC – is the observation that the Bitcoin mining hashrate might grow without any security increments if that growth was slower than Moore's law. For Bitcoin security to increase (thus positively impacting the Bitcoin security/utility), this growth needs to outpace Moore's Law.

According to our calculations, the Bitcoin hashrate needs to grow over 41 %/year to effectively increase the network security and its usefulness, as well as revert a positive net mining increment from the miners.

Although, from a certain point, adding more hashrate to the network has a very marginal increase in network security in the short term, continuously adding hashrate to the network is important for long-term network maintenance, resilience, and security.

If Bitcoin's hashrate falls behind Moore's law for a prolonged period, this can threaten Bitcoin's long-term utility and security.

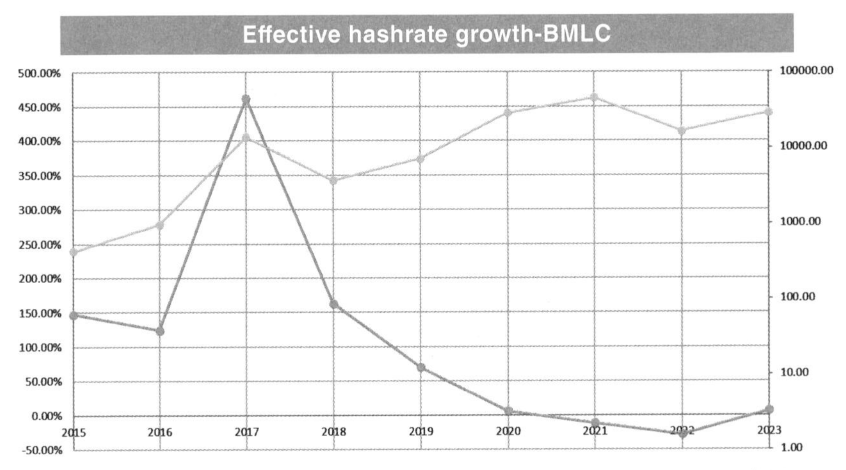

Chart: Effective hashrate growth according to BMLC (Deep Green) and the Bitcoin price (Light green). Log scale.

The chart above displays the Bitcoin price on a logarithmic scale and the effective hashrate growth considering BLMC. According to this, after discounting for Moore's law, the hashrate growth has been negative for 2021, 2022, and 2023.

For BMLC, we have used the following:

BMLC = ([end year total BTC hashrate –
start year total BTC hashrate]/
[start year total BTC hashrate]) – 41%

The Moore's law states that the microchip power doubles approximately every two years. This can be represented by a power of 2. In terms of percentage growth, this doubling represents about a 41% increase per year since 2^(1/2) = √2, which is approximately 1.41. Although this is an approximation and the actual growth rate may vary, we use the -41% in the model.

Saying this, we can still find a positive correlation between price and hashrate.

To find and measure this correlation, we started by plotting the yearly hashrate delta with the yearly price delta, which can be seen in the chart below.

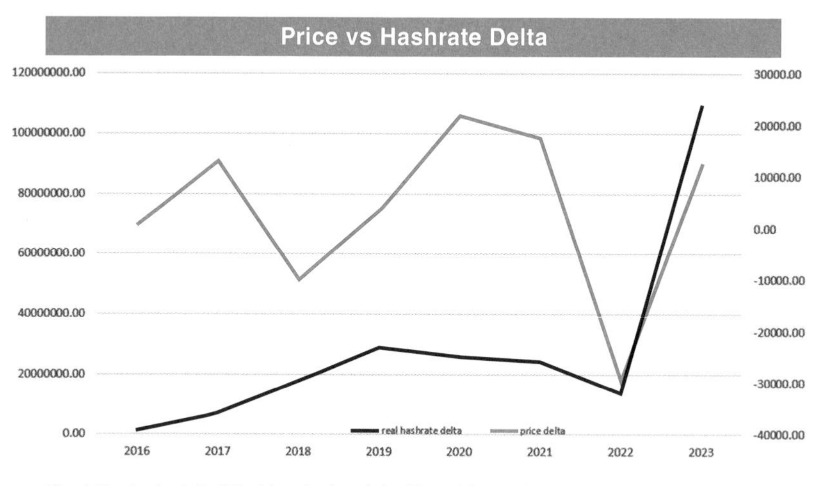

Chart: Hashrate delta (Black) and price delta (Green) for each year period.

The chart is evidence that, indeed, price and hashrate are correlated, but they sometimes suffer deviations. According to our calculations, the BMLC here represented a real hashrate delta, and the price is highly correlated with an R^2 of 0.92.

We can now create a ratio dividing the price delta by the hashrate delta.

Price/hashrate delta ratio = [period price delta]/[period hashrate delta]

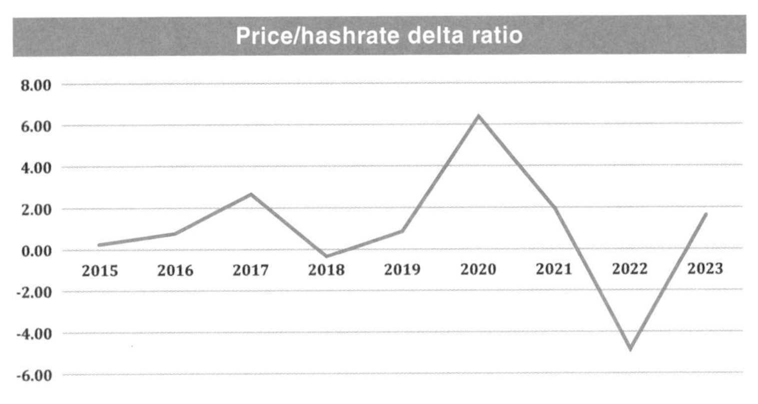

Chart: Price to hashrate delta ratio.

This chart helps us understand how much price and hashrate changed in proportion, thus allowing us to understand the impact of the price of hashrate and vice-versa.

We can read that whenever the ratio is positive, it means that the price increases more than the hashrate, and this translates into the fact that Bitcoin is more expensive in comparison to hashrate variations.

The Price/hashrate delta ratio can be a useful indicator to assess if the price is overvalued or undervalued in correlation to the hashrate.

As we can see on the chart, it is very patent in 2020 and 2022. The Bitcoin price at the end of 2020 could have been considered overpriced in relation to the hashrate delta, while in 2022 when the Bitcoin price bottomed at a 2-year low at $15k, the ratio was negative and consequently considered oversold.

When the ratio is negative or close to zero, it means that the price growth was negative when compared to the hashrate growth. This might indicate that Bitcoin is underpriced, considering the hashrate variation.

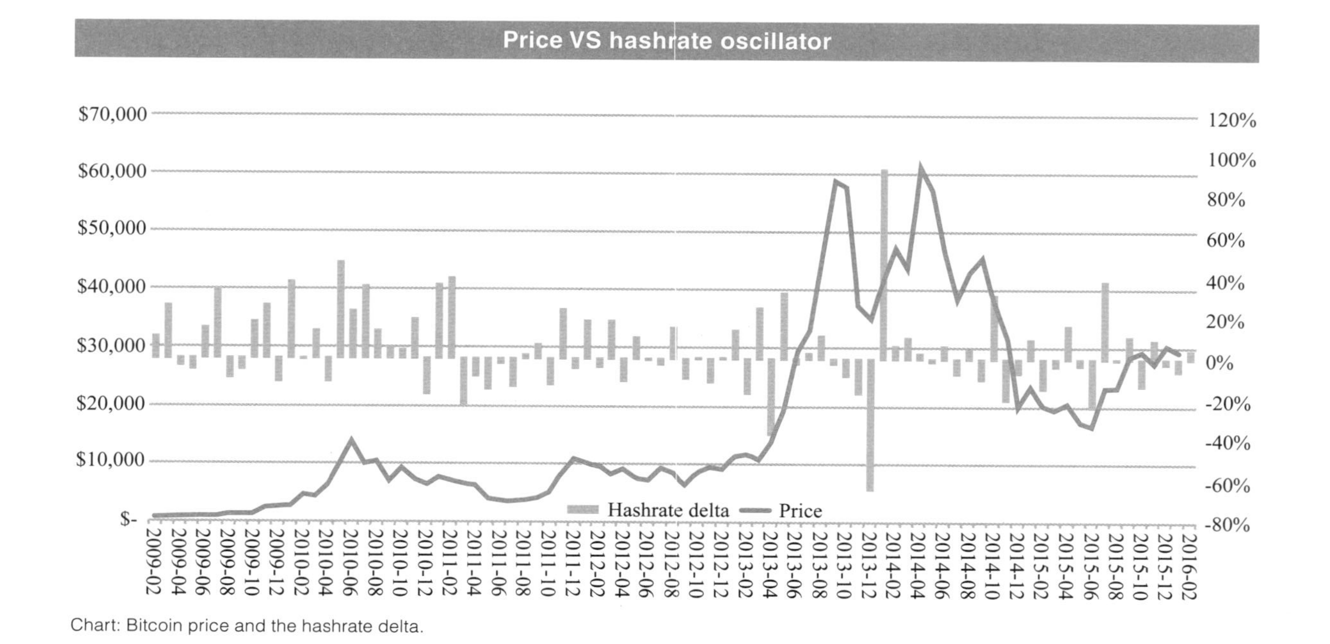

Chart: Bitcoin price and the hashrate delta.

The chart above shows monthly hashrate deltas, which we can easily correlate with price trends. Consequently, it's fair to conclude that the Bitcoin hashrate is a useful tool for predicting price movements and being part of the Bitcoin valuation framework.

The correlation between price and hashrate delta with BMLC can be applied to different time frames, depending on the time horizon that investors are looking for.

Conclusion

The exploration of the relationship between Bitcoin's price and its hashrate, especially when adjusted for Moore's Law, provides a significant understanding of the inherent dynamics of this decentralised network. The Bitcoin Moore's Law Compensation (BMLC) introduces a nuanced perspective on the security and utility of the Bitcoin network. By requiring the hashrate to outpace Moore's Law, it sets a tangible target for the network's growth to ensure its increasing security.

Despite occasional deviations, the positive correlation between Bitcoin's price and hashrate further emphasises their intertwined relationship. Particularly, the Price/Hashrate delta ratio offers a valuable tool for gauging if Bitcoin is overvalued or undervalued relative to its hashrate variation, providing useful insights for investors.

Lastly, it's worth noting that while the findings presented here hold as of the current analysis, the dynamics of Bitcoin and the broader crypto space are ever-evolving. Continuous monitoring and reassessment of these relationships are necessary to stay updated in this fast-paced space. The flexibility to adapt to these changes is inherent to the success of long-term investment strategies in the realm of cryptocurrencies.

1.5 Bitcoin price equilibrium adjusted to M2

In this section, we will use a Bitcoin price equilibrium valuation model to calculate the price according to the equilibrium between supply and demand, adjusted according to the M2 money supply.

The Bitcoin supply is fairly easy to calculate, considering its hard cap, programmed mining inflation rate, and on-chain data that allows us to evaluate the amount of Bitcoin available for trade. On the other hand, the Bitcoin demand might be, at first sight, more challenging to calculate as there are many variables that might influence it. However, according to our research, most of the demand for Bitcoin can be explained by looking at its scarcity (limited supply), the fact that Bitcoin has consolidated its position in diversified portfolios, and the fact that it is highly correlated to the M2 money supply.

Let's break down these concepts in order to understand this price equilibrium valuation methodology better.

Bitcoin supply is very inelastic, but demand has no cap.

An important consideration is that, unlike many other

markets, the Bitcoin supply is not elastic, and there are few product substitutes (if any).

Common examples of elasticity: if the demand for product A goes up, the price will go up, and typically, suppliers will ramp up the production, thus increasing the supply of that product in order to keep up with the increase in demand and price.

Sometimes, consumers can also choose a substitute product: In the automotive market, if prices of new SUVs go up, buyers can get a new sedan. If the sedan is expensive, they can buy a second-hand car. If cars are too expensive, they can buy a motorcycle. Car manufacturers can also produce more cars to meet the demand.

In the crypto space, when using Bitcoin, users buy a piece of the most secure network in the world. There's no substitute product for it, and there's no second best. Additionally, no matter what the price is, Bitcoin supply is fixed, and miners can never increase the production of Bitcoin.

It's also important to consider the price/demand dynamics of Bitcoin. For most products, there's an inverse correlation between price and demand. If the price goes up, the demand declines.

However, Bitcoin demand behaves slightly differently. Buyers can always continue to buy Bitcoin, just in smaller quantities (fractions of a Bitcoin).

Other products with inelastic demand are, for example, gasoline. When the price of gasoline goes up, the quantity purchased isn't greatly affected because people continue to need it for transportation.

The same happens to Bitcoin. Bitcoin can continue to provide the exact same benefits no matter what the Bitcoin price is. Bitcoin is the fuel for the most secure computer network in the world. Users who find Bitcoin useful will continue to use it. Bitcoin will continue to have demand as a reserve asset, disregarding its price. Bitcoin's usefulness for users is the same at a $10,000 price tag or at a $100,000 price tag.

Comparisons can also be made with gold. Gold is hard to mine, and the increase in annual supply is usually between 1% and 2%. Even if the price of gold increases by 30%, the change in quantity (i.e., inflation) of gold will continue to be very small, considering that gold continues to be very hard to mine. This means that the quantity/supply of gold is very inelastic. This is very different from other common metals like bronze or copper. If the price of copper increases by 30%, the supply will most likely also increase significantly.

Additionally, just like Bitcoin, gold continues to have the same chemical properties and offers the same use cases at $500 per once or at $2000 per once.

Regarding the Gold supply, according to GoldHub, even in years where the gold price increases significantly, like in 2020, where the gold price increased by 24%, the gold supply (mine production) increased only by 1.5%.

What does this say about the gold elasticity?

- Gold elasticity of supply coefficient = % change in quantity/ % change in price
- X=1.5%/24%
- X=0.0625
- It's very inelastic

When a good has a coefficient less than 1, it is considered to have an inelastic supply. Gold's supply elasticity coefficient is considered very low, at 0.0625.

Here's how we can calculate Bitcoin's elasticity of supply:

Bitcoin elasticity of supply coefficient = % change in quantity/% change in price

In Bitcoin's case, the elasticity of supply coefficient is extremely small, much smaller than gold's elasticity of supply. In fact, Bitcoin's elasticity of supply is 625 times smaller than gold, reaching less than 0.0001 in some years. In other words, one could say that Bitcoin is 625 times scarcer than gold.

Additionally, Bitcoin's supply declines every time there's a block halving. This means that Bitcoin's elasticity of supply coefficient will tend to be even smaller over time.

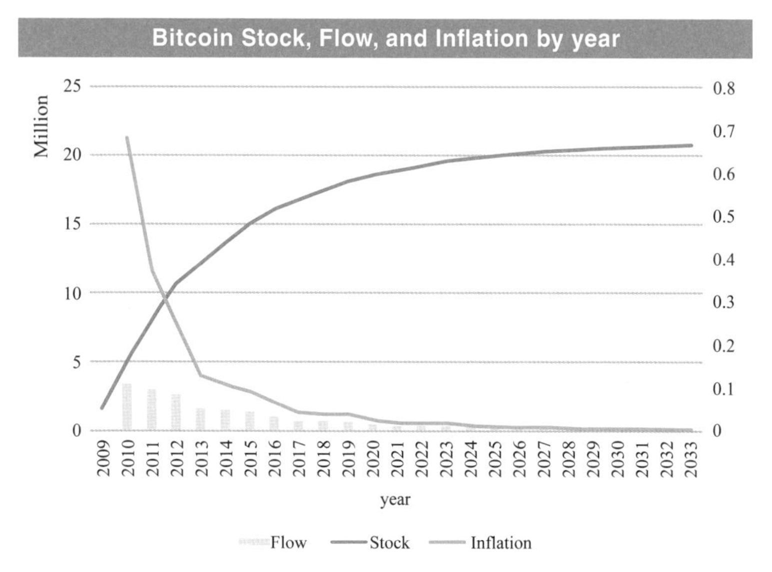

Chart: Bitcoin's flow, stock, and inflation rate.

The chart above shows the total Bitcoin supply as "stock," the Bitcoin "flow," and the Bitcoin inflation rate.

It is very apparent in this chart that, considering the slowing flow and slowing inflation, Bitcoin's elasticity of supply will continue to decrease.

The very low Bitcoin price elasticity of demand that we observe means that Bitcoin price is highly sensitive to changes in demand.

In this context, it is also useful to talk about Bitcoin's income elasticity. Income elasticity of demand measures the responsiveness of the quantity demanded for a good/service to a change in income.

Income elasticity = % change in quantity demanded/
% change in income

In Bitcoin's case, when income increases (which we can measure through the M2 money supply), Bitcoin demand increases (while supply remains the same).

Products with income elasticity greater than one are considered superior goods. The demand for these products increases when the price increases. On the other hand, products with a negative income elasticity are considered inferior goods, and the higher the income, the less the demand for them.

Bitcoin elasticity = bitcoin income elasticity of demand
[(final income-initial income)/(final
bitcoin demand-initial bitcoin demand)]

This measures how sensitive the demand for Bitcoin is to a change in income. Income, in the case of Bitcoin, is highly correlated with the money supply in the market.

Bitcoin's 10-year price projection is based on supply/demand price equilibrium laws and the M2 money supply.

A lot of people in the finance and crypto industry believe that the Bitcoin price will reach $1 million. Cathie Wood, Michael Saylor, Balaji, Jimmy Song, Larry Lapard, and many more all think that the market dynamics will appreciate Bitcoin in the long term.

In this section, we will lay down the math on why and when it would be possible for the Bitcoin price to hit $1 million. We will start with the assumptions.

1) Bitcoin's elasticity of supply coefficient is very low.

   As we already saw before, Bitcoin's elasticity of supply and stock-to-flow ratio is even lower than gold. While gold supply has been very stable over the last 100 years, increasing 1% to 2% every year, Bitcoin's supply is programmatically defined, and it will continue to decline. The table below shows the Bitcoin's supply inflation rate from 2020 to 2025.

2020	2.51%
2021	1.78%
2022	1.75%
2023	1.73%
2024	1.13%
2025	0.84%

   The only thing that increases when the Bitcoin price increase is the Bitcoin mining profitability, consequently attracting more miners to the network, which increases the hashrate, which can be a measure of the network's security.

2) Bitcoin's income elasticity is very high. Similar to luxury goods, Veblen goods, and rare metals (like gold), the elasticity is >1, meaning that the quantity demanded changes more than proportionately to a change in income. People are more likely to buy Bitcoin when they have more income available to them.

   Income elasticity = % change in quantity demanded/ % change in income

3) M2 Money supply: M2 represents all the money in circulation. It includes bank deposits, cash, cheques, etc. and it is provided by the FED and Central Banks around the world.

The M2 money supply increases mainly for two reasons: GDP growth and Central Banks issuing more currency.

4) Bitcoin has a very positive correlation with the M2 money supply for three main reasons:

    - Supply elasticity: Bitcoin's supply is fixed and very inelastic. No matter how much Bitcoin's price increases, the supply remains the same.
    - Income elasticity: an increase in M2 money supply is correlated to the increase of nominal income of individuals and businesses.
    - Inflation: an increase in M2 money supply often correlates with inflation, and Bitcoin is seen as a "store of value" or "digital gold," protecting investors against inflation.
    - Asset allocation: an increase in circulating money increases the flow to investable assets. As of 2022, the global AUM was $131 trillion. The crypto market cap is now at $1.3 trillion, which corresponds roughly to 1% of the world's total AUM. Bitcoin's market cap is roughly 50% of that; in other words, it is around 0.5% of the world's AUM.

All these factors create demand for Bitcoin that can only be met with a price increase.

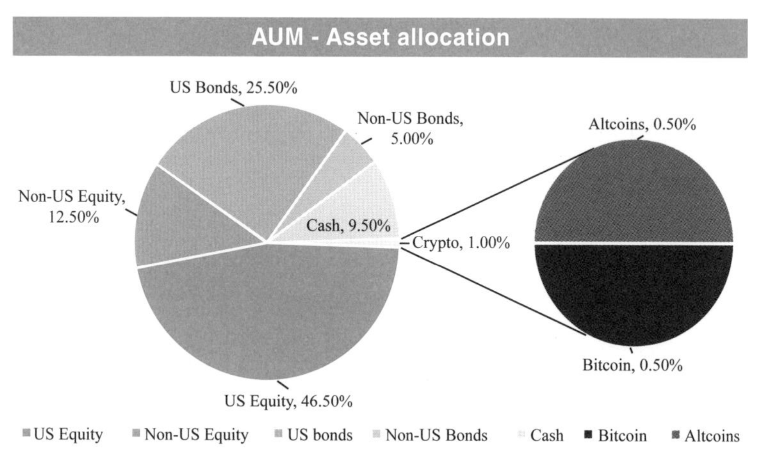

Chart: World's AUM distribution in 2023.

The chart above represents the world's asset management asset distribution as of Feb. 2023. According to Bitcoin's market capitalisation, Bitcoin should represent 0.5% of all investable assets in the world. Money always looks for a home.

Let's go back to the correlation between the M2 money supply and investable assets, in this case, the S&P 500:

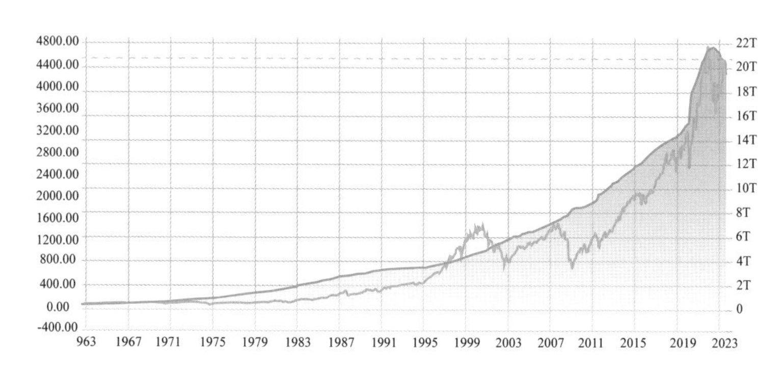

Chart: The S&P 500 index (Light green) and the US M2 Money Supply (Deep green). 1963 to 2023. Source: TradingView. Federal Reserve data.

This chart illustrates well the correlation (and, in fact, causation) between the M2 supply and the S&P 500. The chart goes back to 1960, and it's possible to infer that part of the money circulating in the economy will end up flowing to the S&P 500.

An important difference between the S&P 500 and Bitcoin is that companies that are part of the index will issue more stocks and somehow dilute their supply, while Bitcoin, as we saw before, can never increase its supply.

Although we can already detect a meaningful correlation just by looking at the charts, we have decided to run some regressions to better understand the correlation between M2 and the S&P 500. The conclusions are extremely clear.

After plotting the S&P 500 and the M2 for the last 15 years, we have observed that both are highly correlated. However, applying some lags would give us an even higher correlation. The M2 at an 8-month lag gives us an r^2=0.98. This idea was originally explored by Man Yin To on SeekingAlpha.

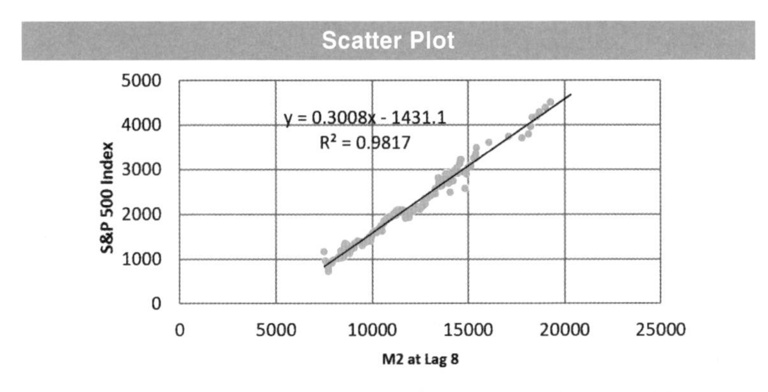

Chart: Regression plotting the S&P 500 and the US M2 money supply at 8-month lag.

Saying this, it is possible to predict with very high accuracy the S&P 500 price:

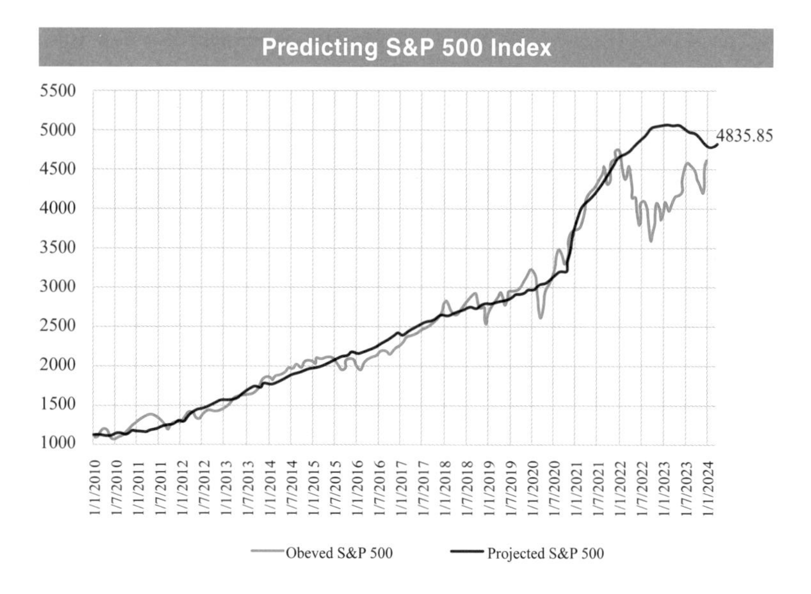

We have also observed that Bitcoin's correlation with M2 is 0.86, and the correlation with the S&P 500 (SPX) is 0.90.

These values give very high statistical relevance to it, as the M2 can be a great predictor of demand for multiple financial products, including Bitcoin, and it should be included in Bitcoin valuation models.

Different indicators tell us that the M2 money supply will continue increasing in the long run. From historical patterns to the fact that governments need to "print money" to pay interest on their debt and fund government activities, we can say with a very high degree of confidence that most countries will continue inflating their money supply. The only solutions to get out of this Central Bank inflationary need would be to either back all currencies with a hard asset (gold standard or Bitcoin standard) or stop printing money and declare bankruptcy. Knowing that these two solutions are highly unlikely, the only future one can foresee is a continuous M2 supply increase, which reaffirms the assumptions for this model.

Note, however, that the M2 supply might, in the short term, contract. Governments might adopt Quantitative Tightening (QT) policies in order to control inflation, as we have seen in 2022 and 2023. This will temporarily contract the money supply in the economy, which, paired with higher interest rates, might also translate into a contraction in the flow of money to investable assets.

However, we are confident that the M2 money supply will continue increasing, validating the valuation model used in this Bitcoin price equilibrium.

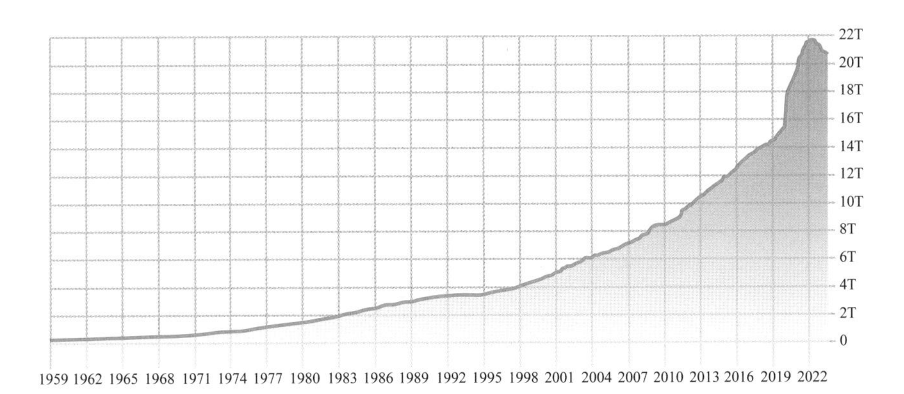

Chart: The US M2 Money Supply.
Source: TradingView. Federal Reserve data.

What we can conclude from the previous chart is that the money supply in the market will continue trending up in the medium/long term as long as we live in a world riddled with monetary policies where politicians believe that the fairy-tale of Keynesian monetarism works.

Although we will focus on the US M2 supply, not only the US M2 but also China, Japan, India, Russia, Canada, and the EU have similar monetary policies. All countries that use fiat currencies will end up printing money as much as or even more than the US in order to keep up with world trade and USD reserves. This phenomenon is called the Dollar Milkshake Theory.

Predicting Bitcoin's price through the M2 Money Supply

As we saw before, the 10-year correlation coefficient between M2 and the S&P 500 is 97%. We can observe this by plotting both on a chart and measuring their standard deviations, as we saw in the previous pages.

Now, let's look at some additional assumptions for this model.

In this price equilibrium model, we account for the fact that a percentage of Bitcoin holders are long-term holders, effectively removing a large amount of Bitcoin from circulation. In addition to that, according to some estimations, 30% of all Bitcoin has been lost forever.

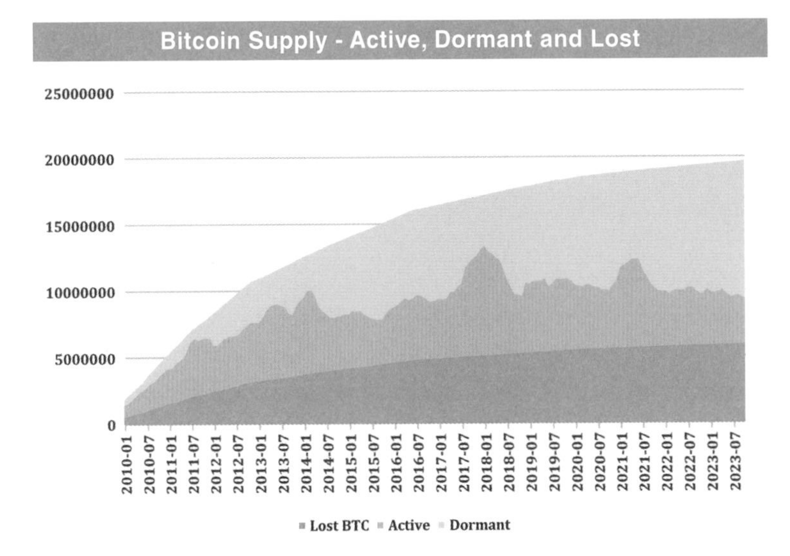

Chart: Bitcoin supply over time.

This chart illustrates well how the total Bitcoin supply is broken down in terms of active circulation, dormant BTC, and Lost BTC.

Long-term holders (dormant on the chart) are considered Bitcoin holders who hold BTC for over 1-year. The trend continues to increase.

The fact that the real circulating supply is low and most holders are long-term holders makes Bitcoin's velocity low. However, if we take into consideration the velocity of Bitcoin's M0 and M1 alone, the velocity is way higher than the overall M2 and M3 velocity.

We can say that:

M0 = Speculative Bitcoin
M1 = Bitcoin held in exchanges, hot wallets, wBTC.
M2 = M1 + cold storage, Bitcoin hodlers, Bitcoin locked in DeFi, etc.
M3 = Bitcoin lost forever

The following three charts also add weight to the efficacy of our Bitcoin equilibria model. The first shows a non-trivial correlation between Bitcoin and inflation. The second shows us that the correlation coefficient between Bitcoin and M2 is 0.86. The third shows us a correlation between Bitcoin and the S&P 500 index (SPX) of 0.90.

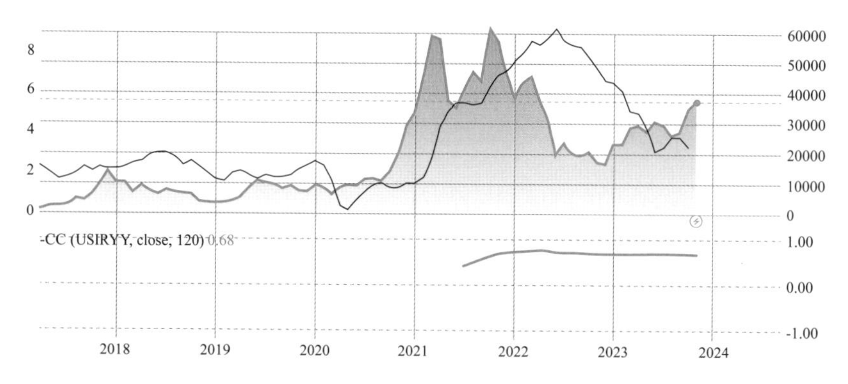

Chart: Bitcoin price and the US inflation rate have a 10-year correlation coefficient of 0.68.
Source: TradingView, Bitstamp, Department of Labor's Bureau of Labor Statistics

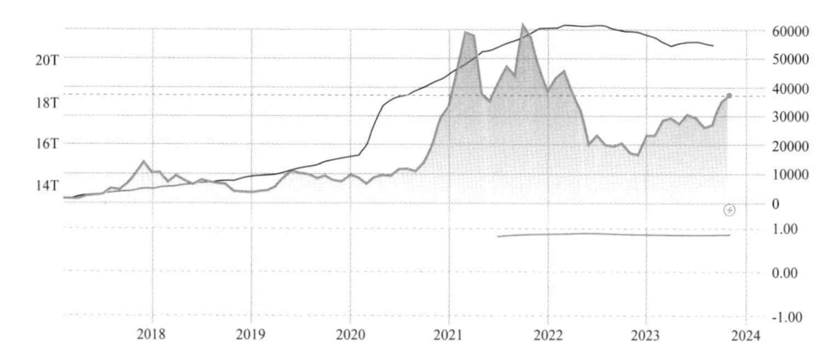

Chart: Bitcoin and the M2 money supply correlation coefficient is 0.86.
Source: TradingView, Bitstamp, Federal Reserve

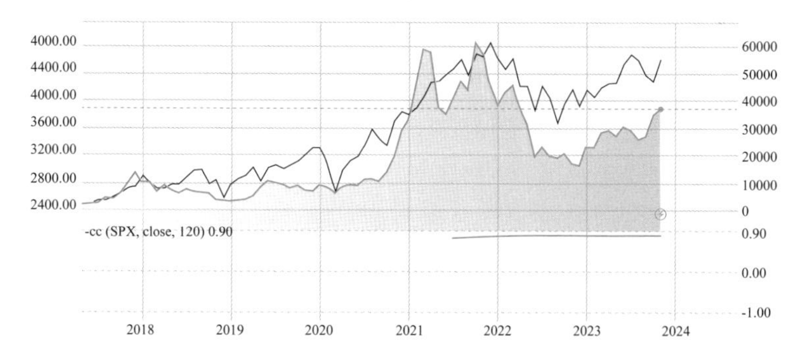

Chart: Bitcoin and SPX correlation is 0.90.
Source: TradingView, Bitstamp, Standard and Poor's.

Considering the previous assumptions and the correlations above, we have created a model to evaluate and predict the Bitcoin price.

These are the assumptions and calculations of the model:

1. Our assumptions start with the M2 money supply. A share of the existing and future money supply flows to investable assets.

2. Money supply is a good predictor of the stock market performance, especially the S&P 500. Although the money supply with no lag has a significant R^2 at 0.94, the M2 at lag 8 months gives the best regression to predict the S&P 500 with an R^2 as high as 0.98.

3. We also found that the S&P 500 is a good Bitcoin predictor: $R^2 = 0.88$

4. This is also aligned with the Bitcoin and M2 correlation coefficient of 0.86 that we saw before, given by the correlation coefficient $r = \Sigma[(xi - X)(yi - Y)]/\mathrm{sqrt}([\Sigma(xi - X)^2][\Sigma(yi - Y)^2])$

5. In July 2023, the Bitcoin market cap is $568 million, which represents approx.. 1.535% of the S&P 500 market cap: approx.. $37 trillion, and 0.5% of the total AUM.

6. The Bitcoin/S&P 500 allocation can be expressed by $y = 0.0009x - 0.0136$, where x is the market cap of the S&P 500. This function is the result of a linear regression that correlated the S&P market cap and the Bitcoin market cap as a percentage of the S&P 500 index. The R^2 was 0.82, giving it a high statistical significance.

7. From here, we infer that the Bitcoin market capitalisation can be predicted by the following:

   $BTCmcap=(0.0009(SPmcap)-0.0136)*SPmcap$

8. Knowing Bitcoin's market capitalisation, we can calculate the Bitcoin price equilibrium. To calculate the price better, we need to consider the quantity of Bitcoin in circulation. For this, we have added a "hodler factor" that can go from 10% to 80%. This is expressed in the following manner:

   $y = MIN(6, MAX(0, (A*\sin(ft+0))))$

Where A is the *hodle* rate amplitude, f is the frequency, and t is the time which corresponds to the Bitcoin price of the previous period. This creates a feedback loop between the price and the *hodle* rate. This *hodle* factor is the variable that accounts for the fact that a percentage of Bitcoin users will not sell BTC, no matter what (including those Bitcoin lost forever).

9. Finally, we can predict the Bitcoin price:

   BTC price = initial price * (1 + ((end spmcap * 0.0009 - 0.0136 – start spmcap * 0.0009 - 0.0136)/(start spmcap * 0.0009 - 0.0136) * ((1/(1-hodlefactor) + MIN(6, MAX(0, (A*sin(ft+0))))

According to our prediction, Bitcoin's price might hit the $1 million mark by late 2032.

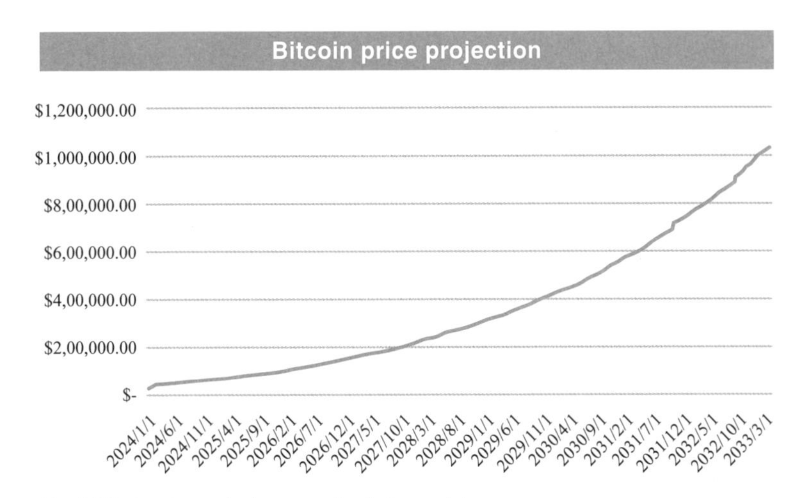

Chart: Bitcoin price projection according to the model.

This theoretical model can also be used for the valuation of the total cryptocurrency market capitalisation, which, as we saw, corresponds to close to 1% of the total AUM.

According to our research, we have verified that the total cryptocurrency market capitalisation has an 84% correlation coefficient with the M2 money supply and a 90% correlation with the S&P 500 index.

Saying this, it is possible to infer that an increase in money supply will increase the flow allocated to crypto assets (which at the moment is around 1%), consequently increasing its demand.

Having the money supply as a guide for the increase in the demand for the crypto market demand increase, investors can use it as a strong predictor of future trends. Adaptations of this model can be used to measure the impact of this demand on the total crypto market.

Additionally, investors looking to understand the impact of the M2 money supply on other assets can do similar calculations according to the weight of the asset in the total crypto market cap. For instance, when writing this book, Bitcoin's dominance (its weight on the total market capitalisation) is approximately 50%, while Ethereum's is 17%.

Roughly speaking, we can assume that Ethereum has 17% of the 1% AUM associated with crypto and that an x increase in M2 would, in that proportion, be converted into demand for Ethereum.

The same model might be useful to evaluate well-established native cryptocurrencies with characteristics similar to Bitcoin.

Conclusion

While the model built here provides a novel way to predict Bitcoin and other cryptocurrency prices, it's crucial to remember that this approach is largely theoretical and relies on a series of assumptions that are heavily reliant on monetary policies.

The model's reliance on the M2 money supply as a predictor of asset prices, although it has shown a significant correlation, may not always hold true in the future. The dynamics of the global economy, regulatory actions, technological advancements, and other unforeseen events could disrupt this correlation.

In addition, the model's assumptions about Bitcoin holders and the amount of Bitcoin considered lost or dormant could also change over time. The 'hodler factor' is particularly difficult to accurately quantify and predict, as it depends on the individual behaviours of countless participants in the Bitcoin market, which a wide variety of factors can influence.

In conclusion, while this model presents a compelling theoretical approach to predicting Bitcoin prices, it should be used cautiously. It's important for investors to understand the assumptions and limitations inherent in this model and to consider a wide range of information and approaches when making investment decisions.

1.6 Bitcoin Stock-to-Flow Model

The stock-to-flow (S2F) model has been used to predict the price of assets such as commodities and precious metals. More recently, PlanB popularised its application to Bitcoin.

The S2F model is based on the concept that the value of an asset increases as its scarcity increases. The 'stock' is the total supply of Bitcoin, and the 'flow' is the rate at which new Bitcoins are created through mining. The flow of Bitcoin is halved approximately every four years in an event known as the 'halving.'

The stock-to-flow model correlates the production and the current stock of an asset. In the case of Bitcoin, the stock is the total available supply available in the market, while the flow refers to the new production that is added to the existing stock. In the case of Bitcoin, this new production corresponds to the new coins mined, which halves every year.

To calculate the Bitcoin S2F, we can start by determining the stock. The stock at any given moment corresponds to the total circulation supply of the coin. To determine the flow, we need to calculate the number of new coins entering circulation for a specific period. In the case of Bitcoin, there's a new block produced every 10 minutes with a 6.25 BTC reward, and this reward is cut by approximately half every four years, thus reducing the flow.

The S2F ratio is the stock (total supply) divided by the flow (new coins)

$$S2F = Stock/Flow$$

In order to use the S2F ratio, we have applied a multiplier that applied to the S2F model, returns a price prediction with a correlation to the historical price of 85%. To find this multiplier, we have used an optimization process to find a number that maximises the value of the correlation coefficient. In this case, the multiplier is 421.8.

S2F price projection = 421.8 * (stock/flow)

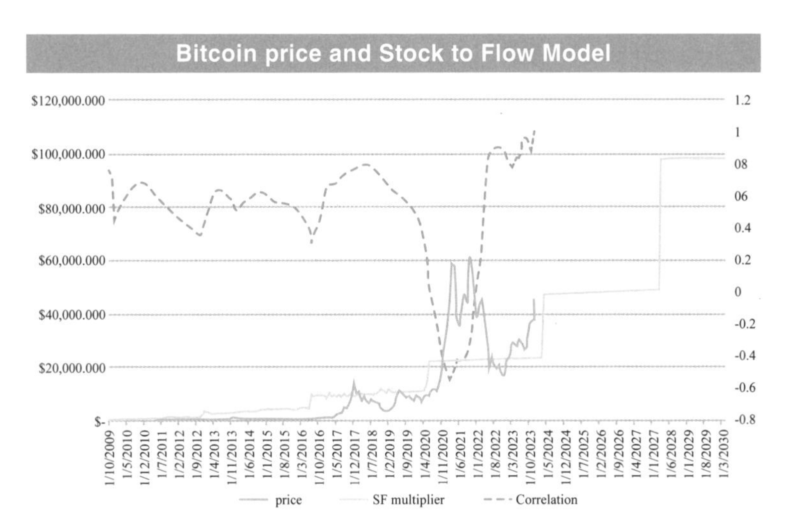

Chart: Stock to flow model in Light green, Bitcoin price in Deep green and correlation between the two in Dotted line (axies on the right).

The stock-to-flow model has been accurate several times in the past, and it is widely used in the financial industry.

The red line on the chart above shows the S2F model and predicts the Bitcoin price, while the line in green represents the correlation coefficient between the S2F model and the Bitcoin price, which we can see is highly correlated most of the time.

One of the criticisms of the S2F model is that it doesn't account for the demand. However, one could argue that the demand is intrinsic to the price and consequently might not need to be included in the mode.

As of Jan. 2024, the total number of Bitcoins mined was approximately 19.6 million. Given the current block reward of 3.125 Bitcoins (post-April 2024 halving), approximately 164,250 new Bitcoins will be mined each year until the next halving. Therefore, the S2F ratio as of 2024 would be approximately 119 (19.6 million/164,250), which is notably higher than the ratio for assets such as gold and silver.

Using the multiplier provided, the S2F price projection for Bitcoin would be

S2F price projection = 421.8 * (stock/flow)
S2F price projection = 421.8 * (19.6 million/164,250)
S2F price projection = 421.8 * 119
Price = $50k

Critics of the S2F model argue that it oversimplifies the dynamics of the Bitcoin market, as the S2F model does not account for demand. While some argue that demand is reflected in the price, others point out that demand can be influenced by a wide range of factors, including regulatory changes, technological advances, market sentiment, and economic conditions. These factors can cause significant deviations from the price predicted by the S2F model.

Additionally, the S2F model assumes that the halving of the block reward will always lead to a significant increase in price. However, this may not always be the case.

Conclusion

In conclusion, the S2F model is a valuable tool for understanding the potential impact of Bitcoin's scarcity on its price. However, like all models, it is not infallible, and it should be used in conjunction with other analysis methods and market information when making investment decisions. The S2F model doesn't account for the demand and focuses mostly on the supply side, which can be limiting by not accounting for a good part of the reasons why prices move.

1.7 Bitcoin P/S ratios

P/S ratios and P/E ratios are very commonly used in the stock market for stock valuation and as an indicator that correlates the current price of a stock with its revenue or earnings.

The P/S ratio allows investors to evaluate how the market values every dollar that the company generates. It is commonly used to evaluate companies (especially growth stocks) that don't have a profit yet. Looking at the P/S ratio makes it possible to understand if the asset is undervalued or overvalued when compared to other companies in the same sector and compared to its historical performance.

Although Bitcoin doesn't have revenue according to the traditional accounting definition of revenue, Bitcoin does have revenue if we consider the block rewards and transaction fees as revenue. Additionally, similarly to many growth stocks, Bitcoin doesn't have earnings or profits.

The Bitcoin network itself doesn't have expenses either, as the network itself runs free of charge. Although Bitcoin miners do need to spend money on hardware and electricity, we can consider this an externality, considering that the Bitcoin network, the Bitcoin price, and respective functionalities are not correlated to Bitcoin miner costs - unlike traditional business where the cost of developing a product is "internalised."

Saying this, we can come up with a crypto P/S ratio by simply looking at the transaction fees, block rewards and the Bitcoin circulating supply. This ratio can be used as a valuation tool and as a tool to compare the prices of a cryptocurrency to its earnings.

Note also that calculating the P/S of a DeFi protocol can resemble a traditional business. The P/S ratio can also be a useful tool for the valuation of DeFi protocols, as DeFi protocols do sell a service and generate fees.

However, in this case, we can also take a look at the Bitcoin P/S, as the Bitcoin network generates fees and block rewards for the miners. As we will see, the Bitcoin P/S ratio chart resembles the S2F chart. This is because the total miner revenue used to calculate the P/S ratio is directly linked to the Bitcoin flow.

P/S = Market capitalisation/Sales
P/S = Crypto price/(total miner revenue/circulating supply)
P/S = Crypto price/((fees + block rewards)/circ. Supply)
P/S = 30 000/(($832,211,538.46)/19 460 110)
Bitcoin P/S = 665.3

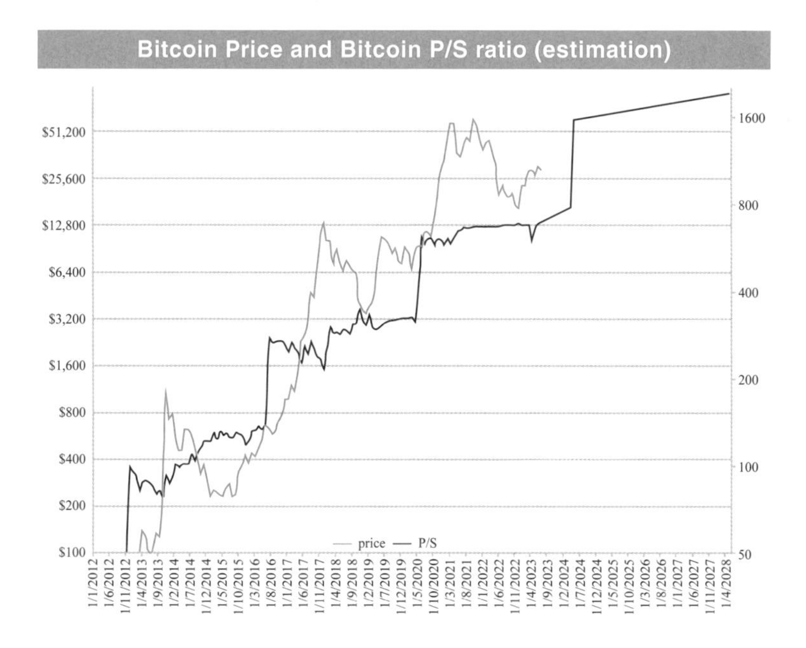

The projection above takes Bitcoin's P/S ratio into consideration. In some aspects, this is similar to the stock-to-flow model. However, we have identified the recursive effect between price, miners' rewards, and their impact on the P/S ratio. Then, according to the future block rewards and the 2024 block halving, we extrapolate what the price target for Bitcoin would be at a certain P/S ratio. This is *ceteris paribus*, of course.

Although the P/S ratio is heavily influenced by the Bitcoin supply (and consequently the stock to flow), we include other variables too.

This method is very simple, but it can help us understand how much the price deviates from the Bitcoin P/S ratio and conclude if the price is overvalued or undervalued relative to the ratio.

Conclusion

The P/S ratio can be useful not only for Bitcoin but also for other cryptocurrencies, especially layer 1 chains. However, this mode does have some limitations.

The P/S ratio ignores the fact that the issuance of new coins to reward miners/validators might cause sell pressure and compromise the feasibility of the project and the network in the long term. This should not be the case with Bitcoin, as BTC issuance decreases over time.

The P/S ratio might not be ideal for comparing different cryptocurrencies with substantially different fee structures. For instance, Solana has a very different "sales" model, and consequently, it would be unfair to compare Bitcoin's P/S with Solana's P/S.

Overall, while the P/S ratio can provide interesting insights into the value of Bitcoin and potentially other cryptocurrencies, it should be used in conjunction with other metrics and analysis methods to get a more complete picture of a cryptocurrency's value.

1.8 Bitcoin comparison with gold ETF

Bitcoin has historically been considered similar to gold as a store of value and as a hedge against inflation. Both assets are seen as a store of value with limited supply, independent from traditional financial systems and with global recognition. Considering this, we can speculate on how a Bitcoin spot ETF will affect the future price of Bitcoin by comparing it to the relationship between the AUM and the price of gold ETFs.

According to research from the World Gold Council, the consensus in the market is that the global economy may experience a short-term recession, and the interest rate hikes in developed countries such as Europe and the United States may come to an end. However, historically, a mild economic recession and a cooling inflation coupled with a weakening dollar can usually be favourable factors for gold. In fact, we can observe that the Dollar Index (DXY) and the price of gold are typically inversely correlated. Similarly, Bitcoin is also inversely correlated with the DXY.

With the added support from geopolitical conflicts, gold as a hedging tool, its price performance may be further supported in the future as "digital gold." In addition to being affected by macro factors and the narrative of halving every four years, Bitcoin is likely to be stable due to its decreased monetary policy uncertainty.

1.8.1 Bitcoin - "Digital Gold"

Bitcoin has long been considered a value store similar to gold and a hedge asset. The CEO of BlackRock, Lary Fink, Cathie

Wood of ARK, Bank of America, Deutsche Bank, JPMorgan, Citigroup, Goldman Sachs, Morgan Stanley, Fidelity, and other well-known institutions have all publicly expressed similar views comparing Bitcoin to gold. The analogy between Bitcoin and gold is mainly based on the following aspects:

Scarcity: Like gold, the Bitcoin supply is limited. The total supply of Bitcoin is fixed at 21 million, and its supply is capped. Gold is also a scarce resource, with a limited supply that is difficult to increase (although it's actually easier to mine gold and increase its supply over time).

Value Storage: Gold is widely regarded as a storage asset with inflation-resistant and value-preserving properties. To some extent, Bitcoin has also been considered to have a storage function because its supply is limited and not subject to government intervention. Bitcoin's design makes it resistant to inflation.

Market Sentiment: Gold is often seen as a hedge asset in the market. When investors' economic uncertainty increases or market volatility rises, they tend to invest in gold. To some extent, Bitcoin is also regarded as a hedge asset because it is relatively independent of traditional financial markets and is sometimes seen as a tool to hedge inflation and political instability in some cases.

Market Behaviour: Both gold and Bitcoin attract the attention of speculators and investors, and their price fluctuations are affected by market demand and psychological factors. Both can be used as investment tools to pursue capital appreciation.

Therefore, we can explore the future price trend of Bitcoin by analogy with gold.

1.8.2 The Institutionalisation of the Gold Market

The World Gold Council points out that the development of gold trends and the gold market structure had banks playing a leading role in it, while non-bank institutions also gradually became important liquidity providers for gold, including hedge funds, high-frequency trading companies, etc., retail investors and wholesale markets, represented by ETFs, are also increasingly interacting, thereby affecting gold prices.

When financial institutions issue gold ETFs, it may have an impact on the gold price for the following reasons:

Increased Market Liquidity: Issuing gold ETFs can provide investors with a convenient way to invest in gold because ETFs can easily be bought and sold on exchanges. Such a move might increase the liquidity of the gold market, attracting more investors to participate in the gold market and thereby affecting the gold price.

Increased Investment Demand: The launch of gold ETFs may stimulate investors' interest in gold. Investors can gain exposure to gold by buying shares of gold ETFs without having to hold and store physical gold. Therefore, the issuance of gold ETFs may increase investment demand for gold by lowering the barriers to exposure to gold.

Market Sentiment and Expectations: The issuance of gold ETFs may affect market sentiment and expectations. Suppose investors are optimistic about the prospects of the gold market, such as an increase in hedge demand, rising inflation expectations, and lower market interest rates. In that case, they may increase demand for gold ETFs that give investors a convenient way to invest in gold.

The table below shows the top 20 gold ETFs by AUM as of November 6, 2023. The two largest ones are GLD, issued by State Street Global Advisors (SSGA) in 2004, and IAU, issued by BlackRock's iShares in 2005, with asset management sizes of $55.6 billion and $25.8 billion, respectively. Our analysis below mainly selects these two ETFs as representatives.

Symbol	ETF Name	Issuer	Inception Date	Total Assets ($M)
GLD	SPDR Gold Shares	State Street Global Advisors	Nov 18, 2004	55,611.50
IAU	iShares Gold Trust	BlackRock	Jan 21, 2005	25,778.80
GLDM	SPDR Gold MiniShares Trust	State Street Global Advisors	Jun 25, 2018	5,979.80
SGOL	Abrdn Physical Gold Shares ETF	Abrdn Plc	Sep 09, 2009	2,731.97
BAR	GraniteShares Gold Shares	GraniteShares	Aug 31, 2017	937.97

Symbol	ETF Name	Issuer	Inception Date	Total Assets ($M)
IAUM	iShares Gold Trust Micro ETF of Benef Interest	BlackRock	Jun 15, 2021	931.84
OUNZ	VanEck Merk Gold Trust	Van Eck	May 16, 2014	754.12
AAAU	Goldman Sachs Physical Gold ETF	Goldman Sachs	Jul 26, 2018	547.75
UGL	ProShares Ultra Gold	ProShares	Dec 01, 2008	173.59
DBP	Invesco DB Precious Metals Fund	Invesco	Jan 05, 2007	133.81
DGP	DB Gold Double Long Exchange Traded Notes	Deutsche Bank	Feb 27, 2008	79.25
IGLD	FT Cboe Vest Gold Strategy Target Income ETF	First Trust	Mar 02, 2021	79.12
FGDL	Franklin Responsibly Sourced Gold ETF	Franklin Templeton	Jun 30, 2022	69.30
IAUF	iShares Gold Strategy ETF	BlackRock	Jun 06, 2018	53.59
BGLD	FT Cboe Vest Gold Strategy Quarterly Buffer ETF	First Trust	Jan 20, 2021	28.08

Symbol	ETF Name	Issuer	Inception Date	Total Assets ($M)
GLL	ProShares UltraShort Gold	ProShares	Dec 01, 2008	11.21
DZZ	DB Gold Double Short Exchange Traded Notes	Deutsche Bank	Feb 27, 2008	4.42
SESG	Sprott ESG Gold ETF	Sprott	Aug 02, 2022	3.97
DGZ	DB Gold Short Exchange Traded Notes	Deutsche Bank	Feb 27, 2008	3.19
GLDX	USCF Gold Strategy Plus Income Fund ETF	Marygold	Nov 03, 2021	3.15

Source: VettaFi, Compiled HashKey Capital

1.8.3 The Impact of Gold ETFs on Gold Prices

The following chart shows the correlation between the AUM of GLD and IAU and the spot price of gold. The R-squared values of both regressions exceed 0.9. Therefore, for GLD and IAU, the two largest gold ETFs in terms of asset size, there is still a strong correlation between the changes in the spot price of gold and changes in the corresponding ETF's AUM.

We selected the past ten years of data (from August 30, 2013, to August 16, 2023) for GLD and IAU's AUM and the corresponding gold spot time range.

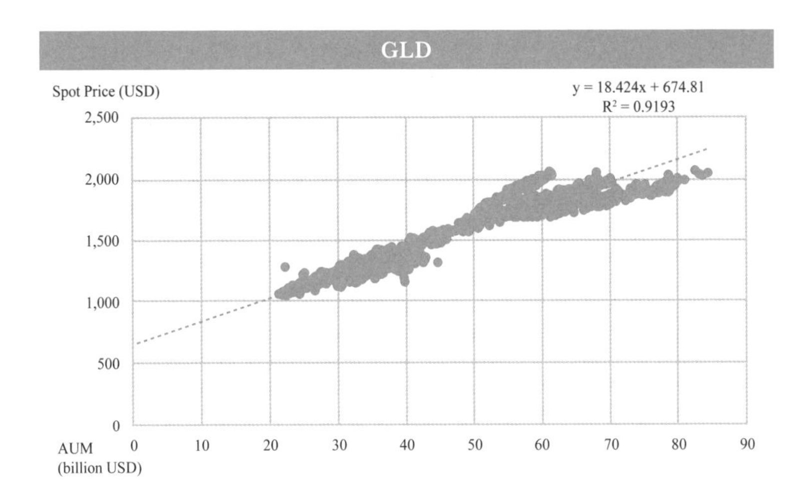
GLD
Spot Price (USD)
y = 18.424x + 674.81
R² = 0.9193
2,500
2,000
1,500
1,000
500
0
AUM
(billion USD)
0
10
20
30
40
50
60
70
80
90

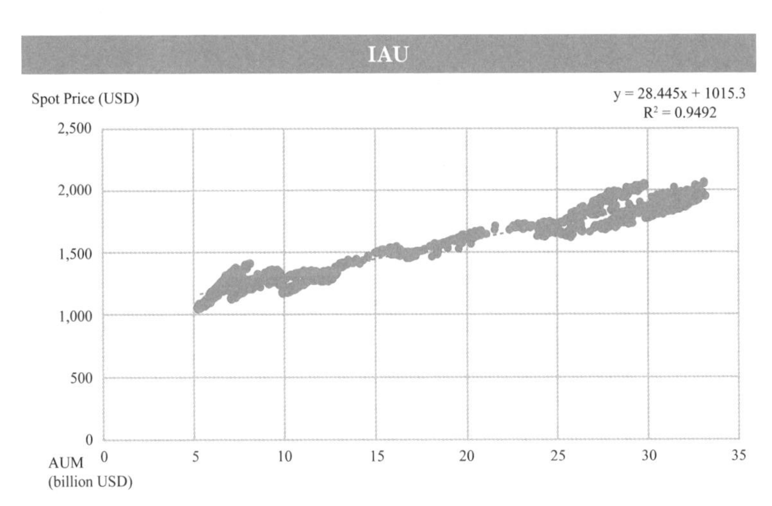
IAU
Spot Price (USD)
y = 28.445x + 1015.3
R² = 0.9492
2,500
2,000
1,500
1,000
500
0
AUM
(billion USD)
0
5
10
15
20
25
30
35

A rough observation of the change in gold prices after the issuance of ETFs also reveals the impact of ETFs on prices: We use different time windows of one day, two days, one week, two weeks, one month, two months, half a year, and one year to study the changes in gold prices after the issuance of GLD and IAU. Especially in shorter time windows, such as one day, two days, one week, two weeks, and one month, the price of gold is mostly rising in these periods. However, the change in gold prices is influenced by a variety of factors, including global economic conditions, geopolitics, inflation expectations, market interest rates, monetary policy, etc.

	Inception Date	GLD	IAU
		Nov 18, 2004	Jan 21, 2005
1 day	ETF Transaction Volume (USD)	11,655,300	759,500
	Gold Price change%	0.80%	0.95%
2 days	ETF Transaction Volume	23,651,300	1,107,000
	Gold Price change%	1.30%	0.28%
1 week	ETF Transaction Volume	36,023,300	4,509,000
	Gold Price change%	2.07%	0.82%
2 weeks	ETF Transaction Volume	56,353,100	7,486,500
	Gold Price change%	2.76%	-2.10%
1 month	ETF Transaction Volume	86,905,300	11,016,000
	Gold Price change%	0.01%	0.90%
2 months	ETF Transaction Volume	128,644,900	19,080,500
	Gold Price change%	-4.58%	2.22%

	Inception Date	GLD	IAU
		Nov 18, 2004	Jan 21, 2005
6 months	ETF Transaction Volume	276,295,700	37,732,500
	Gold Price change%	-5.03%	2.81%
1 year	ETF Transaction Volume	507,003,900	108,245,500
	Gold Price change%	9.92%	34.00%

Data source: Yahoo Finance, arranged by HashKey Capital

The market value of gold was approximately $12.9 trillion USD on November 9, 2023, while Bitcoin was $713 billion USD. The market value of gold is 18 times that of Bitcoin, so under the same fund inflow, the inflow of funds into Bitcoin ETFs will have a much greater impact on the price of Bitcoin than the impact of gold ETFs on the price of gold.

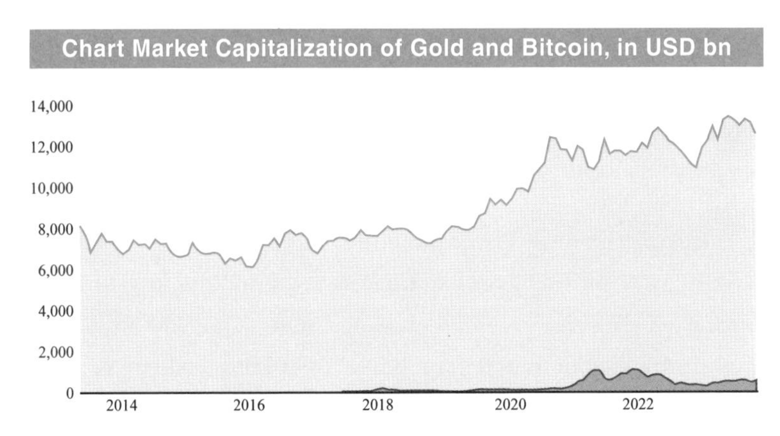

Chart Market Capitalization of Gold and Bitcoin, in USD bn

Data source: Ingoldwetrust

1.8.4 Bitcoin ETFs

Without going into too much detail on the benefits of the Bitcoin ETF as an investment vehicle, some of the benefits might include:

Diversified Portfolio: Cryptocurrency ETFs often can eventually include multiple different cryptocurrency assets, such as Bitcoin, Ethereum, and other major cryptocurrencies. By investing in a cryptocurrency ETF, one can gain exposure to multiple cryptocurrencies, thereby achieving portfolio diversification and mitigating the risk of specific cryptocurrencies. ETFs also open doors for institutional investors to gain exposure to Bitcoin easily.

Simplified Transactions: Compared to directly buying and holding multiple cryptocurrencies, investing in a cryptocurrency ETF can provide a more simplified way of trading. It has a lower threshold for investors, who can participate in the cryptocurrency market by buying or selling ETF shares without having to handle individual cryptocurrency transactions and storage. The ETF is managed by a professional fund team, and investors don't need to keep custody of the crypto assets.

Liquidity and Transparency: Cryptocurrency ETFs are usually listed on exchanges and have good liquidity. Investors can buy or sell ETF shares on the exchange at any time without having to wait for specific market conditions or trading counterparts. In addition, the net asset value and portfolio holdings of ETFs are generally publicly disclosed, allowing investors to understand their investments better.

Wider distribution: ETFs are distributed across large traditional brokerage firms, while cryptocurrencies are typically available only in crypto exchanges.

Regulatory Compliance: Cryptocurrency ETFs typically operate under a regulatory framework, such as that of the SEC. This means they need to comply with specific provisions and disclosure requirements, providing investors with a certain level of regulatory protection.

Multiple crypto ETFs were approved around the world prior to 2024, but its impact has been low due to the fact that they were issued in small markets with small distribution, as well as the fact that many of these ETFs are futures ETFs and not spot (i.e., they hold a Bitcoins contract rather than the asset itself).

The approval of Bitcoin spot ETFs in the United States is crucial because the US stock market accounts for nearly 42% of the total global market value. Additionally, ETFs registered in the US benefit from global distribution. Compared to other countries that have issued cryptocurrency ETFs, such as Canada, Europe, Brazil, and Dubai, the US has a very large stock market value and liquidity.

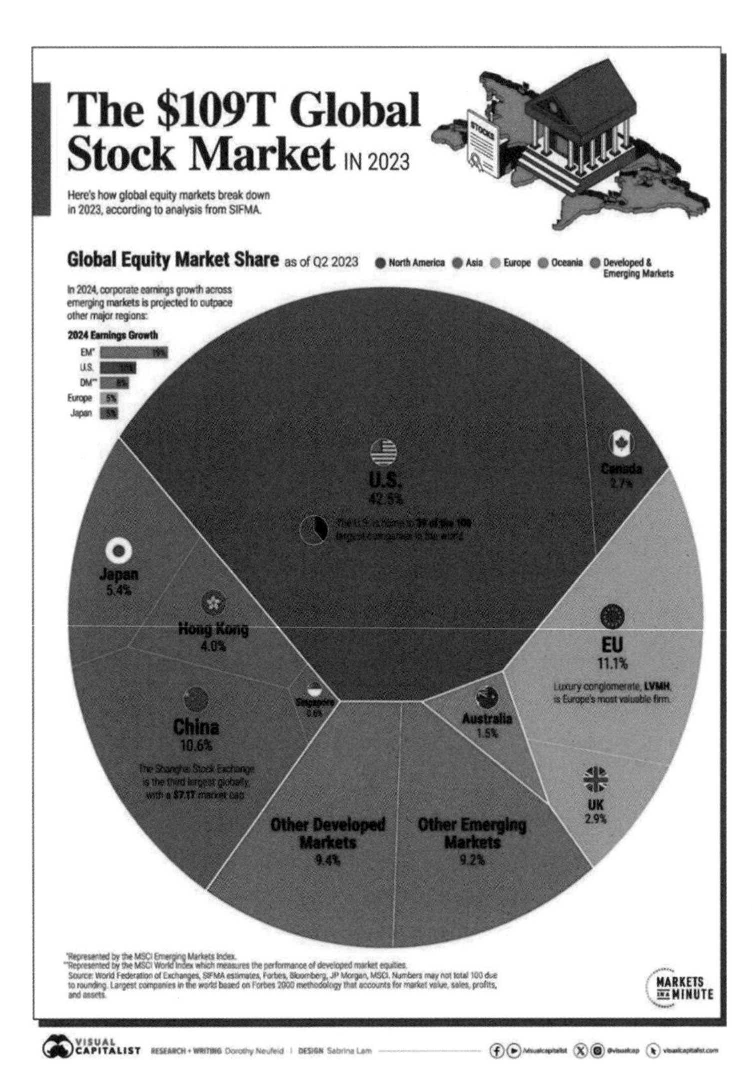

Source: Visual Capitalist

Before the approval of Bitcoin spot ETFs, the United States had already issued Bitcoin futures ETFs, such as ProShares Bitcoin Strategy ETF, Amplify Transformational Data Sharing ETF, Bitwise Crypto Industry Innovators ETF, Global X Blockchain & Bitcoin Strategy ETF, First Trust Indxx Innovative Transaction & Process ETF, etc. Among them, ProShares Bitcoin Strategy ETF is the first Bitcoin futures ETF to be launched in the United States, with assets under management of approximately 1 billion USD.

So, what's the difference between Bitcoin spot ETFs and futures ETFs? The operation of Bitcoin futures ETFs involves the fund management company purchasing Bitcoin futures contracts and creating an ETF based on them. These futures contracts are financial derivatives related to the Bitcoin price, and they do not require the actual holding of Bitcoin, so their impact on the Bitcoin price is relatively low. The net asset value of futures ETFs will be affected by the price of Bitcoin futures contracts. The biggest difference between Bitcoin futures ETFs and spot ETFs is the underlying assets: Bitcoin futures ETFs invest in Bitcoin futures contracts, while Bitcoin spot ETFs invest in actual Bitcoin. Therefore, Bitcoin spot ETFs require the fund management company to purchase an equivalent amount of Bitcoin as a reserve for issuance. This will have a significant impact on the capital inflow, demand, and price in the entire crypto market. This is why the application and approval of Bitcoin spot ETFs in the United States has attracted so much attention.

Finally, on January 11, 2024, the SEC approved the listing and trading of 11 Bitcoin spot ETFs, most of which are operated by well-known asset management companies.

Proposed Bitcoin ETF & management fees as of Jan 10, 2024

Name	Ticker	Issuer	Fee (after waiver)	Waiver Details	Exchange
ARK21Shares Bitcoin ETF	ARKB	ARK Invest & 21 Shares	0% (0.21%)	6 months or $1 billion	CBOE
Bitwise Bitcoin ETP Trust	BITB	Bitwise	0% (0.20%)	6 months or $1 billion	NYSE
Fidelity Wise Origin Bitcoin Trust	FBTC	Fidelity	0% (0.25%)	Until July 31, 2024	CBOE
Franklin Bitcoin ETF	EZBC	Franklin	0.29%	n/a	CBOE
Grayscale Bitcoin Trust (conversion)	GBTC	Grayscale	1.50%	n/a	NYSE
Hashdex Bitcoin ETF	DEFI	Hashdex	0.90%	n/a	NYSE
Invesco Galaxy Bitcoin ETF	BTCO	Invesco & Galaxy	0% (0.39%)	6 months or $5 billion	CBOE
iShares Bitcoin Trust	IBIT	BlackRock	0.12% (0.25%)	12 months or $5 billion	Nasdaq
Valkyrie Bitcoin Fund	BRRR	Valkyrie	0% (0.49%)	3 months	Nasdaq
VanEck Bitcoin Trust	HODL	VanEck	0.25%	n/a	CBOE
WisdomTree Bitcoin Trust	BTCW	WisdomTree	0% (0.3%)	6 months or $1 billion	CBOE

Data Source: Bloomberg

Inflows & BTC held for each ETF									
		Flow (million USD)							
Ticker	Issuer	Day 1	Day 2	Day 3	Day 4	Day 5	Day 6	Total	BTC Held
FBTC	Fidelity	227.0	195.3	102.0	358.1	177.9	222.3	1282.6	30384.0
IBIT	iShares	111.7	386.0	212.7	371.4	145.5	201.5	1428.8	33706.0
BTCO	Invesco	17.4	28.4	31.9	57.6	58.8	63.4	257.5	6192.8
ARKB	ARK Invest & 21 Shares	65.3	39.8	122.3	50.3	41.8	62.6	382.1	9134.2
BITB	Bitwise	237.9	17.2	50.0	68.2	20.1	56.7	450.1	10235.3
HODL	VanEck	10.6	7.3	7.3	4.8	2.3	14.2	46.5	2566.9
BTCW	WisdomTree	1.0	0.0	0.0	1.6	0.0	2.9	5.5	182.1
BRRR	Valkyrie	0.0	19.9	15.0	15.1	1.2	0.0	51.2	1726.5
EZBC	Franklin	50.1	0.0	0.0	1.2	0.0	0.0	51.3	1169.5

Data Source: Bloomberg

1.8.5 Bitcoin price projection from the Bitcoin ETFs outlook

We can make the analogy regarding the potential relationship between Bitcoin spot ETFs and Bitcoin prices from a perspective similar to gold to explore the potential impact of Bitcoin ETF inflows on the price of Bitcoin. According to NYDIG's methodology, assuming a money multiplier of 10 (from November 1980 to July 2023, the average reserve requirement imposed by the FED in the United States was 10.0%, updated monthly, with a total of 513 results. The highest value appeared in March 1992 at 12.0%, and the lowest was 0.0%, according to CEIC Data.

We calculate the money multiplier according to the formula 1/ reserve requirement ratio. This means that an inflow of 1 USD into the ETF will result in a 10 USD impact on the Bitcoin price.

BTC price = ((1/FED reserve requirement ratio) * (ETF Inflow)) + base price

According to the assumption made by NYDIG, the estimated minimum Bitcoin spot ETF AUM is the same as the existing futures BITO ETF, i.e., 1 billion USD, and the maximum exceeds the sum of the AUM of the two gold ETFs GLD and IAU, about 100 billion USD. Combined with the current number of Bitcoins in circulation (BTC outstanding), we can make a prediction about the price of Bitcoin:

ETF Inflow ($M)	Currency Multiplier	Increase in BTC Market capitalisation ($M)	BTCMarket Liquidity (M, 10 Nov)	BTC price variation ($)	BTC Base price ($, 10 Nov)	BTC Price Projections ($)
1,000	10	10,000	19.54	511.77	36,494	37,005.77
2,000	10	20,000	19.54	1,023.54	36,494	37,517.54
3,000	10	30,000	19.54	1,535.31	36,494	38,029.31
4,000	10	40,000	19.54	2,047.08	36,494	38,541.08
5,000	10	50,000	19.54	2,558.85	36,494	39,052.85
6,000	10	60,000	19.54	3,070.62	36,494	39,564.62
7,000	10	70,000	19.54	3,582.40	36,494	40,076.40
8,000	10	80,000	19.54	4,094.17	36,494	40,588.17
9,000	10	90,000	19.54	4,605.94	36,494	41,099.94
10,000	10	100,000	19.54	5,117.71	36,494	41,611.71
16,950	10	169,500	19.54	8,674.51	36,494	45,168.51

ETF Inflow ($M)	Currency Multiplier	Increase in BTC Market capitalisation ($M)	BTCMarket Liquidity (M, 10 Nov)	BTC price variation ($)	BTC Base price ($, 10 Nov)	BTC Price Projections ($)
20,000	10	200,000	19.54	10,235.41	36,494	46,729.41
30,000	10	300,000	19.54	15,353.12	36,494	51,847.12
40,000	10	400,000	19.54	20,470.83	36,494	56,964.83
50,000	10	500,000	19.54	25,588.54	36,494	62,082.54
60,000	10	600,000	19.54	30,706.24	36,494	67,200.24
70,000	10	700,000	19.54	35,823.95	36,494	72,317.95
80,000	10	800,000	19.54	40,941.66	36,494	77,435.66
90,000	10	900,000	19.54	46,059.37	36,494	82,553.37
100,000	10	1,000,000	19.54	51,177.07	36,494	87,671.07

Data source: Coinmarketcap, HashKey Capital

Estimated with the potential future Bitcoin spot ETF's AUM between 1 billion and 100 billion, and all Bitcoins are circulating in the market, the price of Bitcoin will be between 37,005 USD and 87,671 USD.

If we compare this with GBTC's current AUM of 25.17 billion USD (as of December 4, 2023), if the inflow of funds into Bitcoin ETFs reaches the same figure, the price of Bitcoin will reach around $47,000 USD.

However, given the fact that a Bitcoin ETF will be more liquid and with a broader worldwide distribution, as well as accessible to both retail and institutions, it is very likely that the inflow of funds from the American spot Bitcoin ETFs will be much higher.

In addition, although the total supply of Bitcoin is known, not all mined Bitcoins are in circulation. The actual number of Bitcoins circulating in the market is significantly lower than the total supply. Accordingly, based on different hold rate assumptions, the Bitcoin price predicted by this method could be higher.

Conclusion

The impact of ETFs in the gold market can help us draw a comparison with the Bitcoin market and how much the inflow of demand can impact the Bitcoin price.

Investors should be aware of anything that can bring more liquidity to an asset. Typically, when a small cryptocurrency gets listed on a big exchange, the price surges because it means new demand for it. In the same way, a Bitcoin spot ETF being approved is akin to Bitcoin being listed on the biggest traditional exchanges in the world. This is likely to bring important demand inflows that can cause Bitcoin's price equilibrium to change.

1.9 Cost of production method

Can the cost of production be brought into the equation and be used for Bitcoin valuation?

In traditional businesses, the producer of a product can decide at what price he wants to sell the product in a given marketplace by calculating its production costs and adding a margin, taking into account the supply from competitors and demand from buyers. That's true for most consumer products but not for all products in the world.

Let's just take the gold example: if the supply of gold is constant, the gold price is purely determined by its demand, and consequently, the cost of mining gold has little or no impact on gold prices. Typically, it's the demand for gold that moves the price up or down, further driving gold miners to spend more or less in their mining operations, depending on the gold price and how attractive it is to mine gold.

Similarly to gold, the number of Bitcoin miners, how much it costs to mine Bitcoin, and the hash rate also might have a limited impact on Bitcoin's price. Being Bitcoin's supply fixed, the hashrate can only be seen as a metric that measures how much Bitcoin miners believe in Bitcoin's future price and how much profitable it is to mine Bitcoin. The hashrate is key to analysing miners' response to the Bitcoin price. The higher the Bitcoin price is, the more profitable the mining activity will be. Additionally, the higher the hashrate, the more secure the network is. However, the hashrate itself alone does not impact the price. Hashrate can only be seen as miners betting on the future of Bitcoin, but it doesn't impact the price directly, just like the difficulty of mining gold won't impact the price. Instead, the price of the asset will drive miners to embrace more difficult/expensive ways of mining as they become profitable. Consequently, the hash rate or other indicators of mining activity or mining costs do not contribute to price equilibria directly. Despite that, we can definitely drive conclusions by looking at the correlation between Bitcoin mining activities and the Bitcoin price, which we will analyse in this section.

What is Bitcoin Mining?

Bitcoin mining refers to the process of creating and verifying Bitcoin transactions through computer calculations. Mining is a key component of the Bitcoin network. Through mining, participants can compete to solve mathematical problems and receive Bitcoin as a reward. The purpose of mining is to confirm and verify transactions, add them to the blockchain, and keep the network secure and resilient.

In the Bitcoin network, miners use dedicated computer hardware to perform complex mathematical calculations. These calculations are known as "proof of work". Miners need to solve a problem, that is, to find a specific hash value that meets certain conditions. This process requires a lot of computing power and energy consumption. When a miner finds a hash value that meets the conditions, they can add the block to the blockchain and receive a certain amount of Bitcoin as a reward. The reward includes both newly mined Bitcoin and transaction fees collected from transactions.

Bitcoin mining difficulty refers to the level of difficulty in solving the proof-of-work problem in the Bitcoin network. Approximately every two weeks, the difficulty of Bitcoin mining is dynamically adjusted to ensure that new blocks are generated approximately every 10 minutes. As more miners join the Bitcoin network, the increase in computing power will lead to faster mining speeds. In order to ensure a new Bitcoin block is created every 10 minutes, the Bitcoin network will adjust the target difficulty based on past mining and block creation speeds. If the mining speed is too fast, the target difficulty will increase; if the mining speed is too slow, the target difficulty will decrease. The adjustment of Bitcoin mining difficulty is determined by a specific algorithm called the "Difficulty Adjustment Algorithm." This algorithm calculates the new target difficulty based on the mining speed over the past 2-week period (or, to be more accurate, over the last 2016 Bitcoin blocks) to maintain the stability and security of the Bitcoin network.

Due to the constant adjustment of the Bitcoin mining difficulty, it is difficult for ordinary individual users to participate in mining through personal computers. Nowadays, Bitcoin mining is mainly done by professional miners and mining pools, who use dedicated hardware and large-scale computing power to solve the increasing mining difficulty.

The following chart shows the Bitcoin mining difficulty table from April 2013. As of November 6, 2023, the difficulty has reached 62.46T, and the Hashrate (i.e., total network power) has reached 418.92Ehash/s, which is 20 times that of January 2018.

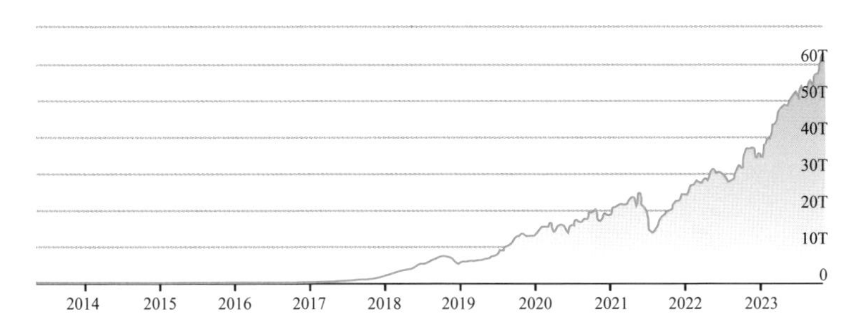

Chart: Bitcoin mining difficulty adjusted over time.
Data source: coinwarz.com

When the market price of Bitcoin is higher than the cost of mining, mining becomes profitable, and miners are incentivised to grow their mining participation. Conversely, if the price of Bitcoin is lower than the cost of mining, miners may face losses and may reduce mining activities or exit the market.

Global Mining Landscape and Cost Analysis

Looking at the regional distribution of computing power worldwide, before China banned Bitcoin mining in June 2021, it provided the largest computing power globally, accounting for more than 50%. This was mainly reliant on the hydropower in Sichuan and the coal power in Xinjiang, with other regions including Yunnan and Inner Mongolia. After the China ban, the United States became the region with the largest share of Bitcoin mining, accounting for about 35%. The chart below also shows the changes in the distribution of computing power by region. According to data from Techopedia, the state of Louisiana has the lowest mining cost in the United States, with one Bitcoin costing $14,955.14. The states with the next lowest mining costs are Idaho, Oklahoma, Wyoming, Utah, and Nevada, while Hawaii has the highest mining cost, with one Bitcoin costing $54,862.05.

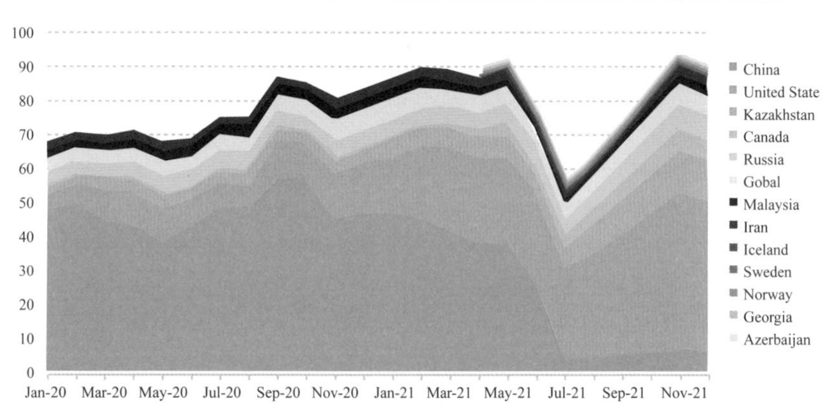

Data source: techopedia.com

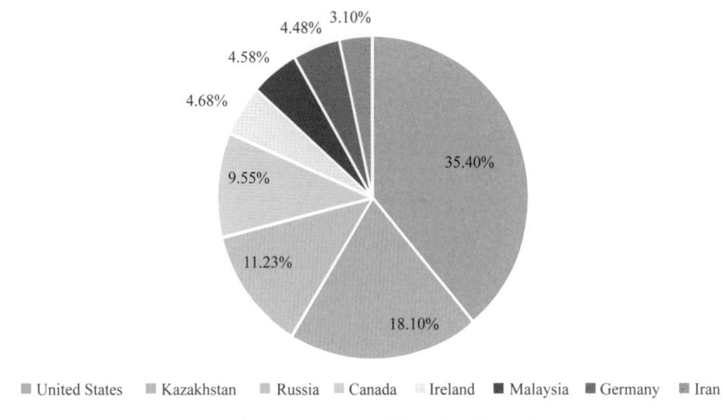

Data source: World Population Review, Compiled HashKey Capital

At the company level, 'Companies Market Cap' summarises the top 18 companies in terms of market value for BTC mining. From the perspective of market value, the largest is Riot from the United States, and the highest revenue is Core Scientific. A small but positive P/E ratio indicates that the company's earnings are higher than its current valuation, suggesting it may be undervalued, while a negative number indicates the company is still experiencing negative returns. The main business of Chinese companies below is the sale of mining machines.

	Name	Symbol	Marketcap	Revenue	PE ratio	Country
1	Riot Blockchain	RIOT	2,262,988,032	256,412,000.00	-6.51	United States
2	Marathon Digital Holdings	MARA	1,990,713,600	174,004,466.00	-2.18	United States
3	Cipher Mining	CIFR	1,028,919,552	56,156,000.00	-82.00	United States
4	CleanSpark	CLSK	708,266,368	141,917,884.00	-2.51	United States
5	Hut 8 Mining	HUT	523,193,088	67,844,173.00	2.68	Canada
6	Bitdeer Technologies Group	BTDR	424,005,344	346,022,000.00	-5.94	Singapore
7	Bitfarms	BITF	326,342,240	82,071,000.00	-1.98	Canada
8	Canaan	CAN	320,749,888	274,634,956.00	-1.51	China
9	Core Scientific	CORZQ	286,575,136	531,389,000.00	-0.33	United States
10	HIVE Blockchain Technologies	HIVE	285,810,784	82,787,252.00	-1.77	Canada

	Name	Symbol	Marketcap	Revenue	PE ratio	Country
11	TeraWulf	WULF	268,007,488	40,420,000.00	-1.53	United States
12	Iris Energy	IREN	234,145,504	46,919,802.00	15.64	Australia
13	Bit Digital	BTBT	204,867,136	38,654,136.00	-2.27	United States
14	Argo Blockchain	ARBK	64,239,004	59,660,646.00	-0.86	United Kingdom
15	BitNile	NILE	44,776,812	107,790,000.00	-0.10	United States
16	Greenidge Generation Holdings	GREE	42,551,540	75,238,000.00	-0.15	United States
17	BIT Mining (500.com)	BTCM	34,448,128	304,967,000.00	-0.21	China
18	Stronghold Digital Mining	SDIG	33,511,130	83,654,550.00	-0.16	United States

Data source: https://companiesmarketcap.com/bitcoin-mining/bitcoin-mining-companies-ranked-by-pe-ratio/

We calculated the breakeven price for several of these companies, divided into the breakeven price counting only COGS (Cost of Goods Sold) and the one counting all costs. The data refers to the official data from the second quarter of 2023 (2023 Q2 financial result) provided by each company. Apart from the number of BTC mined, all units are in US dollars.

	Bitfarms Q2 2023	Hut 8 Q2 2023	Marathon Q2 2023
BTC mined	1,223	905	2,926
COGS	41,519,000	10,756,500	55,222,000
Operation cost	9,155,000	9,404,250	8,546,685
Depreciation & Amortization	9,982,000	7,119,000	37,275,000
Breakeven on COGS	33,948	11,886	18,873
Breakeven on all cost and D&A	49,596	30,143	34,533
	Stronghold Q2 2023	Riot Blockchain Q2 2023	
BTC mined	626	1,775	
COGS	6,291,501	33,842,000	
Operation cost	18,881,835	19,836,000	
Depreciation & Amortization	8,634,767	66,162,000	
Breakeven on COGS	10,050	19,066	
Breakeven on all cost and D&A	54,007	67,515	

Data source: Company official data, compiled by HashKey Capital

Breakeven on COGS = COGS/BTC mined

Breakeven on all cost and D & A = (COGS+Operation cost + Depreciation & Amortisation)/BTC mined

COGS primarily includes costs directly related to the income output of Bitcoin mining activities, such as electricity, premises, machinery, etc. In contrast, Depreciation & Amortisation costs and Operation costs pertain to the company as a whole. Therefore, the actual Depreciation and Operational cost allocated to Bitcoin mining activities would be somewhat less than the reported figures, correspondingly lowering the break-even price on all costs.

Taking Riot Blockchain as an example, according to the information on their official website, in addition to Riot's own mining operations, Riot currently also provides Bitcoin mining hosting services for institutional clients. Apart from the hosting income, Riot generates revenue by providing engineering and construction services for hosting clients, including income obtained through the manufacture and deployment of immersion cooling technology used for Bitcoin mining. According to the table below (Q4 2022 data), self-mining operations accounted for about 3/5 of the total mining volume in Q4 2022 (with the Hosting portion accounting for 2/5). If we roughly reduce the Depreciation & Amortisation cost and Operation cost using the same proportion as in 2022, Riot's break-even price on all costs would be around $48,136.

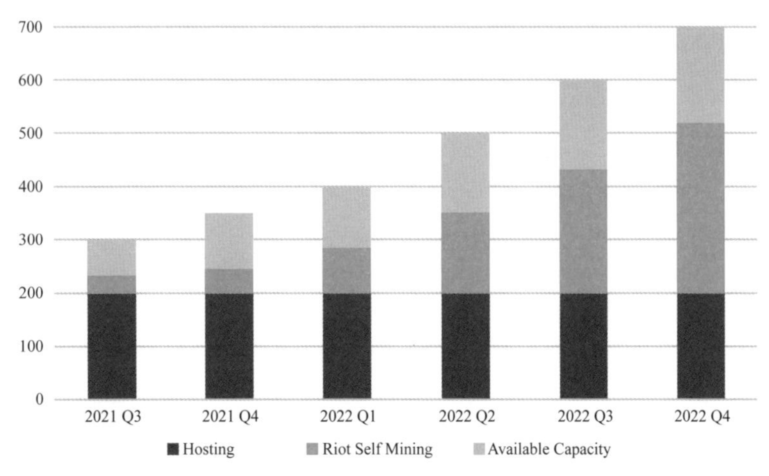

Data source: https://www.riotplatforms.com/bitcoin-mining/whinstone-u-s

If we only consider costs such as electricity and site fees (COGS), the breakeven price for most of the five companies is under 20,000 US dollars. Including all costs, the breakeven price would increase accordingly, varying depending on the company's operating costs.

Mining Machine:

As we mentioned earlier, it is challenging for ordinary users to participate in the mining process with personal computers under the current mining difficulty. Bitcoin mining machines are hardware devices specifically designed for Bitcoin mining, typically composed of specific integrated circuit boards, chips, network cards, fans, etc., with the mining chip (CPU/GPU/ASIC) being the core component that provides computational power. They possess a massive amount of computing capacity and high energy utilisation efficiency, enabling higher mining efficiency in a competitive mining environment. With the development of the Bitcoin network and technological advancements, Bitcoin mining machines are continually evolving. The new generation of mining machines typically has higher computing power and lower energy consumption to adapt to the increasingly competitive mining environment. Among them, personal computer CPUs have no competitive advantage, GPU graphics card mining machines are expensive, and ASIC mining machines are dedicated mining chips, more professional compared to CPUs and GPUs, which can also realise mass production.

From the perspective of mining machines, according to BTC.com data, the Antminer S19XPHydro currently yields the highest return, with a shut-off price of $16,229.18 and electricity expenditure accounting for 46%. The shut-off price of mainstream mining machines is generally in the range of $16,000 to $22,000.

Rank	Miner	Coins	Power Consumption	Computing Power	Daily Output	Break-even Price	Electricity Expenditure	Electricity Cost Rate	Daily Earnings
1	AntMiner S19XPHydro	BTC	5304w	255T	$19.20	$16,229.18	$8.91	46%	$10.29
2	WhatsMiner M53	BTC	6554w	226T	$17.02	$22,627.57	$11.01	65%	$6.01
3	AntMiner S19XPro+Hydro	BTC	5445w	198T	$14.91	$21,464.21	$9.15	61%	$5.76
4	AntMiner S19XP	BTC	3010w	140T	$10.54	$16,787.33	$5.06	48%	$5.48
5	WhatsMiner M33S++	BTC	6820w	220T	$16.57	$24,194.74	$11.46	69%	$5.11
6	Avalon A1366	BTC	3250w	130T	$9.79	$19,507.80	$5.46	56%	$4.33
7	AntMiner S19XPHydro (162T)	BTC	4780w	162T	$12.20	$23,022.88	$8.03	66%	$4.17
8	WhatsMiner M505	BTC	3276w	126T	$9.49	$20,274.55	$5.50	58%	$3.99
9	WhatsMiner M33S+	BTC	6732w	198T	$14.91	$26,531.18	$11.31	76%	$3.60
10	WhatsMiner M505	BTC	3248w	112T	$8.43	$22,642.99	$5.46	65%	$2.97

Data source: BTC.com

Overall, there is a certain relationship between the price of Bitcoin and the cost of mining, but market supply and demand also impact the price of Bitcoin. Theoretically speaking, the price of Bitcoin needs to be higher than the mining cost for miners to have a profit margin. However, the price of Bitcoin is not entirely determined by the cost of mining. The market price of Bitcoin is mainly affected by a combination of factors such as supply and demand, investor sentiment, participation of large institutions, and regulatory policies. The demand for Bitcoin from investors, trading volume, market liquidity, and other factors also have a significant impact on the price, with the mining cost being one of these influencing factors.

Conclusion

From an investor perspective, looking at the mining company's breakeven can help understand if Bitcoin is oversold or overbought. Additionally, Bitcoin mining companies investing in the space show these companies' faith in the crypto market's future and consequently can be one more reading for the market sentiment. Finally, it helps to understand the profitability of the mining companies themselves as a potential alternative investment.

1.10 Transaction value to transaction fee

The Bitcoin network's efficiency, usability, and volume scalability have been improving over time, and this can be measured by dividing the average transaction volume by the transaction fee.

The number of transactions on the Bitcoin network is indeed limited to its block capacity – 1MB – which handles around 4000 transactions per Bitcoin block, translated into seven transactions per second. One could think that the volume could be seriously limited. However, that's not what we observed. Over the years, the average transaction volume has been increasing and peaked in Nov. 2021, when the average size of a Bitcoin transaction was $1.6 million.

During this period, it was very usual to see Bitcoin blocks securing over $10 billion every 10 minutes. This is a remarkable achievement and proof that the Bitcoin blockchain can securely handle very high-volume transactions.

It is important to remember that from the blockchain perspective, the transaction size in dollar value or BTC value doesn't impact the transaction fees at all. Instead, what impacts the transaction fees is the size of the transaction in terms of bytes.

By looking at the transaction value to transaction fee, we aim to:

- Access the utility of the network. The higher the transaction volume when compared to the transaction fees, the more efficient the network is in securing big volumes of transactions.
- Compare how efficient the network is when compared to the Bitcoin price. For this, we will use the ratio "price to volume per $1 fee" outlined below.

Correlating the average transaction size and the network fees

When correlating the average transaction value on Bitcoin and the average transaction fee, we can observe that although the transaction fees might rise in periods of high network demand when looking at the dollar amount of those transactions, the transaction fee is actually decreasing.

In other words, the average cost per transaction might sometimes be higher, but the volume that those transactions carry is also higher.

Saying this, we can argue that the transaction cost on the Bitcoin chain has been not increasing but decreasing.

For example, in 2017, the average transaction value was $27,609, and the average transaction fee was $10.27. if we divide the average transaction value by the average transaction fee, we get that for every $1 fee paid, users moved $6,583.

Fast forward to 2021, and we have observed an average transaction value of $896,222 and an average transaction fee of only $2.90. This means that for every $1 fee paid, the network is moving $309,042, making Bitcoin one of the cheapest ways of moving value across the world.

This is also an important indicator of the usefulness and value of the Bitcoin network.

The results show that, in fact, it's getting cheaper to transact on the Bitcoin network, and more value is moved over the network.

Year Dec.	Average transaction value	Average transaction fee	Volume per $1 fee	BTC Price	Price to volume per $1 fee
2011	$596.00	$0.01	$19,200.00	$3.32	0.00
2012	$673.00	$0.01	$67,300.00	$11.92	0.00
2013	$9,211.00	$0.17	$54,182.35	$383.00	0.01
2014	$3,914.00	$0.06	$69,892.86	$372.00	0.01
2015	$5,084.00	$0.06	$84,733.33	$345.00	0.00
2016	$5,840.00	$0.22	$26,545.45	$702.00	0.03
2017	$67,609.00	$10.27	$6,583.15	$9,170.00	1.39
2018	$20,403.00	$0.47	$43,410.64	$5,443.00	0.13
2019	$25,850.00	$0.70	$36,928.57	$8,060.00	0.22
2020	$116,088.00	$5.78	$20,084.43	$15,622.00	0.78
2021	$896,222.00	$2.90	$309,042.07	$55,830.00	0.18
2022	$128,017.00	$1.20	$106,680.83	$18,200.00	0.17
2023	$58,186.00	$1.80	$32,325.56	$28,000.00	0.87

There are several explanations for these numbers.

- Increase in Bitcoin value
- Growing adoption to move large amounts of money
- Institutional adoption
- Bitcoin use as a store of value
- Growth of the Lightning network for microtransactions

$$\text{Price to volume per \$1 fee} = \text{BTC price} / (\text{avg tx value} / \text{avg tx fee})$$

A low ratio "Price to volume per $1 fee" might translate into the fact that the Bitcoin network is being used for large volumes when compared to the Bitcoin price (i.e. the Bitcoin network cost-effectiveness compared to the price is good), while a high "Price to volume per $1 fee" means that the average transaction size is smaller when compared to the price (i.e. the network cost-effectiveness is lower when compared to the Bitcoin price).

Note that the transaction fee dynamics might change over time as new use cases start to emerge in the Bitcoin chain. These use cases include L2s such as Stacks where DeFi applications can run and Bitcoin Ordinals.

Finally, it is also important to say that Bitcoin is not only a very cost-efficient way of moving value but also very energy-efficient. Although Bitcoin receives some criticism because of its energy consumption, if we compare the energy consumption to other industries (such as banking and gold) and correlate it with the volumes transacted, we can easily conclude that Bitcoin's energy efficiency is amazing when compared to other ways of moving value.

Conclusion

Bitcoin's capacity to handle high-volume transactions securely and efficiently has been demonstrably improving over time, as evidenced by the increasing average transaction volume and the decreasing transaction fees relative to the volume. The confidence of the market is also patent in the very large billion-dollar average transaction volumes seen in the past.

The measure of "price to volume per $1 fee" provides a nuanced perspective on the network's efficiency, revealing that for every $1 fee paid in 2021, the network moved $309,042 - a remarkable increase from $6,583 in 2017.

Several factors contribute to these impressive figures, including the rise in Bitcoin's value, growing adoption for large-scale money transfers, increased institutional adoption, Bitcoin's use as a store of value, and the growth of the Lightning Network for microtransactions.

The low "price to volume per $1 fee" ratio suggests that the Bitcoin network is primarily being used for large volumes. This is also likely to be lower with the Bitcoin ETFs covered earlier in this book. However, this dynamic may evolve as new use cases emerge on the Bitcoin chain, including Layer-2 solutions like Stacks, which facilitate the operation of Decentralized Finance (DeFi) applications, and Bitcoin Ordinals.

We can definitely conclude that Bitcoin is a way more cost-efficient way to move value, being infinitely more cost-efficient than moving gold, as well as way more cost-efficient than moving value in the traditional financial system. We can also add that Bitcoin is very energy efficient considering its use case and volumes transacted.

It's important to remember that these numbers don't signal a cap on the network's potential. On the contrary, they represent a testament to the Bitcoin network's evolving capabilities and scalability. As Bitcoin continues to mature and its ecosystem expands with new technologies and use cases, we can expect the network to become even more efficient and versatile, further enhancing its value and utility.

Chapter 2

Public Chain/ Layer 1 Valuation

In this section of the book, we will dive into a number of valuation methodologies that fit crypto assets related Layer 1 Chains and public chains. In the previous section of this book, "Valuation of Native Cryptocurrencies," despite the fact that we have mostly used Bitcoin as an example, some of the methodologies can also apply to other public chains. However, we find that the methodologies used in this section of the book are more appropriate for valuing public chains and smart contract chains, and investors will more easily find value by applying the methods below.

Method	Goal	Usability
Discounted Cash Flow	Assessing the value of a crypto asset according to its cash flow	Any crypto asset that generates cash flow
Calculating the price equilibrium for Ethereum	Assessing the value of a crypto asset according to its future supply and demand equilibrium	Any crypto asset
Value of a perpetual bond	Estimating the value of a crypto asset according to the yields generated	Crypto assets that generate a real and predictable yield from staking
Metcalfe's Law	Measuring the value of the crypto network according to the network effects generated	Public chains such as L1 and L2 chains
NVT Ratio	Measuring the relationship between the value of a cryptocurrency network and its transaction activity	Public chains such as L1 and L2 chains and some utility tokens
P/S Ratio	Correlating the price of a crypto asset with its revenue	Public chains such as L1 and L2 chains and other fee-generating protocols

What is a public chain?

A public chain, in contrast to a private chain, is a type of distributed blockchain network that allows anyone to participate in a permissionless manner. The public chain is a decentralised network where data and transaction information are jointly maintained and validated by nodes all over the network.

The main feature of a public chain is decentralisation, as it does not rely on the control of a single entity or a central institution but is maintained and validated by multiple nodes distributed across the network. This gives the public chain a decentralised nature, with no single point of failure, and it is more secure and trustworthy.

All transactions and data on the public chain are transparent and can be viewed and verified by anyone. This transparency helps increase trust and reliability, making the public chain a transparent network for value transfer and storage.

Public chains use consensus mechanisms to keep the network secure and to determine which node has the right to add new blocks to the blockchain.

Common consensus mechanisms include Proof of Work, Proof of Stake, and variants of Proof of Stake. Consensus mechanisms ensure the security and consistency of the blockchain.

Decentralised applications

Public chains provide developers with a platform to build decentralised applications (DApps) and smart contracts. Developers can use the smart contract features and Turing-complete programming languages of the public chain to build various blockchain-based applications, such as financial services and digital assets.

Public chains serve as important infrastructure and are the basis for building a good Web3 ecosystem. Public chains are divided into Layer 0, Layer 1, and Layer 2, which we will explain one by one:

Layer 0: Layer 0 is the basic underlying layer that allows interaction and cross-chain communication between blockchains, such as the Pokladot and Cosmos cross-chain ecosystems. Layer 0 can be regarded as the communication or transport layer, solving the problem of blockchain isolation and enabling the transfer of data and assets.

Layer 1: Layer 1 defines the basic rules, consensus mechanisms, data structures, etc., of the blockchain. Layer 1 is the core part of building the entire blockchain system. For example, Bitcoin and Ethereum, which we are familiar with, are representative Layer 1 protocols, each with its own blockchain network and corresponding consensus algorithms.

Layer 2: Layer 2 refers to the protocols and solutions built on top of Layer 1, mainly aimed at improving scalability and performance. The goal of Layer 2 is to offload transactions and data from the main chain to reduce the burden on the main chain and increase throughput and performance. Layer 2 solutions can basically include state channels, rollups, and plasma, among which rollups are the most popular solution. These solutions achieve higher transaction throughput and lower costs by moving transactions and computations off the main chain and submitting the results to the main chain for validation when necessary.

What is the current layout of public chains?

Before discussing the current layout of public chains, let's first understand the blockchain trilemma: The blockchain trilemma refers to the inevitable trade-off relationship between three core components in blockchain technology. These three components are Decentralization, Security, and Scalability.

Decentralisation: One of the core concepts of the blockchain is decentralisation, i.e., the verification and storage of transactions and data without a central institution but rather distributed across a large number of permissionless nodes. Decentralisation can provide higher resistance to censorship, tamper-proofness, and user autonomy, but it also increases the complexity and operating costs of the network.

Security: Blockchains need to have sufficient security to prevent malicious behaviours and attacks. Security includes protecting the integrity of transactions, preventing double-spending and other fraudulent behaviours, and protecting users' assets and privacy. Security and decentralisation often go hand in hand.

Scalability: Scalability refers to the ability of a blockchain network to handle a large number of transactions and data. As the number of users and transactions on the blockchain increases, the network needs to be able to handle more data and transactions without causing congestion and delay.

The concept of blockchain trilemma indicates that in current blockchain technology, it is difficult to optimise these three objectives at the same time. Typically, when one component is strengthened, the other components may be limited. For example, to achieve higher decentralisation, more nodes and wider participation may be required, which could lead to lower scalability. Or, in pursuit of higher scalability and throughput, the degree of decentralisation may be reduced. The development of blockchain technology has been striving to balance these three objectives and has proposed some solutions, such as layered design, improvements to the consensus algorithm, etc. Although Ethereum has tried to solve the trilemma dilemma through multiple upgrades, no public chain has so far been able to balance these three aspects perfectly.

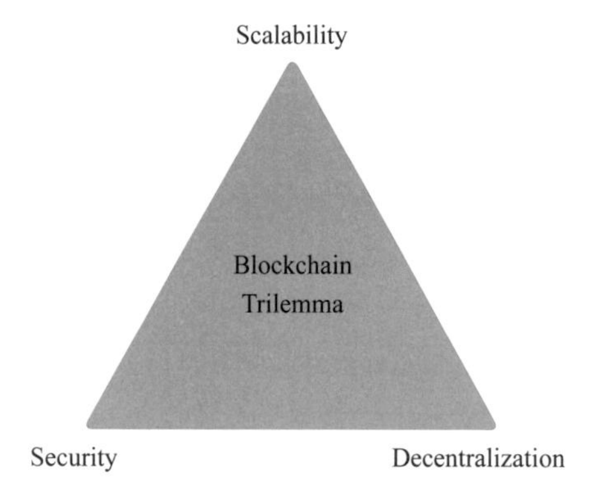

Before 2022, Ethereum held an absolute position in the public chain ecosystem, and consequently, around the EVM-compatible chain, a large ecosystem was developed, such as Avalanche, Tron, BSC, and Layer 2 solutions, such as Arbitrum, Optimism, and zkSync.

Developers can migrate their smart contracts developed and tested on Ethereum to other EVM-compatible chains, thereby obtaining higher performance, lower cost, or other specific functions and advantages. This migration process is usually relatively simple, as the compatibility of code and smart contracts ensures consistent execution on different chains.

Other non-EVM chains include Solana, Cardano, etc., which do not support the Ethereum Virtual Machine and use different virtual machine standards but also might offer different design trade-offs.

According to DeFiLlama data, in January 2021, Ethereum's TVL accounted for more than 95%, and in January 2022, it came to 55.5%. From January 2022 to the present, it has always maintained around 55%. In addition to projects in the previous bull-bear cycle like Solana and Avalanche, there are also some emerging public chains like Sui.

Today, the idea of the public chain is no longer to surpass Ethereum but to create uniqueness in other aspects, such as modular blockchains like Celestia, Move series chains like Sui, app chains (games, DeFi, privacy-dedicated chains) like DeFi native chain Sui, environmentally friendly chains, among others.

Ethereum and L1 valuation

Ethereum, as well as other cryptocurrencies that fuel smart contract blockchains, have characteristics that allow us to evaluate them according to certain metrics, such as volume, number of transactions, fees, and cash flows.

One important caveat when using the DCF method is that we need to apples to apples when it comes to cash flow. As we saw before, different chains address the blockchain trilemma differently. Consequently, it would not be fair to compare, for example, Ethereum to Solana, as the second has way lower fees than Ethereum (but perhaps also lower security).

2.1 DCF – Discounted Cash Flow

Although Ethereum is a decentralised network, it has a revenue source – fees – that are used on the network and partially paid to validators and can be accounted as cash flow.

The legendary investor Warren Buffett has previously expressed that "you can not value the financial asset because it does not produce cash flow. The price is simply based on supply and demand." However, this is not accurate. Many cryptocurrencies and dApps do generate cash flow.

In this section, we will analyse the DCF model to evaluate crypto assets.

Most layer 1 and layer 2 chains have a native cryptocurrency that is used as part of the fee mechanism and to reward node operators (which we interchangeably call miners, validators, or simply nodes).

Fees are important for multiple reasons, but one of the main reasons is security. Blockchains are, by nature, decentralised and highly redundant. This means that a large number of nodes (Ethereum, for example, has close to 1 million nodes) have to synchronise, stay up to date with all the transactions on the network, and correctly validate all the transactions. Fees play an important role in preventing a malicious actor from spamming the network with useless transactions. Fees are also used to reward these network participants.

If there were no fees at all, it would be easy to perform a DoS – Denial of Service - attack on the network by simply adding millions of useless transactions that would clog the network, which would deny the service to legitimate transactions. This has happened in the past to networks such as Solana.

Consequently, fees can be seen as a price to pay to record a transaction on the blockchain in perpetuity. These transactions can be the simple transfer of a coin, the transfer of an asset, the recording of a message, the ownership of an NFT, a governance vote on a DAO, or a smart contract interaction.

Fees are something that can be a fair proxy for the cash flow of the network. Although this kind of cash flow might differ from what we are used to in traditional finance/accounting, it still represents a flow of cash (fees) that are charged by the protocol for users to use it.

There's a caveat to using the DCF model when evaluating a blockchain network. Different chains offer different throughputs, which makes fees differ a lot from chain to chain. For example, a transaction on Ethereum might cost $2, while a transaction on Solana might cost $0.02. Solana will, of course, have a lower cash flow, but at the same time, it has a higher throughput that can benefit certain use cases that value speed over security. Saying this, it makes it hard to compare these two chains' DCF models as they work in fundamentally different manners and offer slightly different services to the users. DCF can only be used to compare a chain with its historical data and should not be used to compare different chains between them.

DCF can also be used to evaluate other crypto projects that generate fees, such as DeFi protocols, L2 chains, bridges, gaming projects, infrastructure projects, prediction markets, oracles, and marketplaces.

DCF Value = ∑ (Estimated Network Revenue in Year N/
(1 + Discount Rate) ^ N) for N = 1 to N years
Per-ETH Value = DCF Value/Current ETH Supply

For the DCF model, we can consider a growth rate for the network, the average cost of capital, risk discount rate, the fact that the node business provides a high-value cash flow with very little overhead. A DECF model can also take into account the base fees plus tips, as well as recognise MEV as part of the revenue. In the future, the cash flow can also account for security as a service (resulting from Ethereum being seen as a Layer 0 and integrating with technologies such as EigenLayer).

Revenue = base fees + miner tips + MEV + security as a service

2.2 Calculating Discounted Cash Flow Method (DCF)

Viewing Ethereum as a Traditional Tech Stock

Ethereum is sometimes viewed as and compared to tech stocks, considering that it provides a tech service – distributed computing – and, market-wise, is seen as a volatile asset with similar behaviour to tech stocks.

As an open, decentralised blockchain platform, Ethereum's core value and growth potential are closely related to the technology and applications it supports.

First, Ethereum is not just a digital currency but a global open-source platform that supports smart contracts and Dapps. Smart contracts make Ethereum distinct from many other blockchains, including Bitcoin. Smart contracts can execute transactions automatically without the need for a third party (such as a legal or financial institution). This innovative technology has led to the rise of DeFi and has found applications in multiple industries.

Second, Ethereum is the foundation of many emerging blockchain projects and crypto assets, including a large number of ERC-20 tokens and emerging assets like NFTs. This means that when these projects and assets succeed, Ethereum may also benefit. In conclusion, the value and potential growth of ETH are closely related to its underlying technological innovation and the applications it supports.

This makes it somewhat similar to tech stocks, as the value of tech stocks also typically depends on the innovativeness of their technology and the breadth of their applications. However, as a crypto asset, the price volatility of ETH may far exceed most traditional tech stocks.

In fact, we have observed that ETH has a correlation coefficient with the Nasdaq index of 0.73, but when we compare it with leverage Nasdaq ETFs, the result is even more noteworthy: the correlation with QLD – (ProShares Ultra QQQ a 2x leveraged ETF) is 0.88, and the correlation with TQQQ (ProShares UltraPro QQQ a 3x leveraged ETF) is an impressive 0.94. Similar results can be observed when plugged into S&P 500 data.

Here, we use the Discounted Cash Flow (DCF) model to value ETH. The DCF model is based on future cash flow expectations to estimate the current value of a stock. Compared to technical analysis that focuses only on market prices, DCF better reflects the true value and long-term investment value of a company or asset.

This can be extremely useful to investors.

Determine the discount rate (r) and growth rate(g)

The formula for the DCF model is as follows:

$$DCF = \sum (CF\_t/(1+r)^\wedge t)$$

Where:

CFt = Cash flow in the t-th year;
r = Discount rate (expected rate of return of the investment)
CFt = (1+g)t-1 CFt-1
g = Growth rate

When using the DCF model for valuation, the terminal growth rate is typically set to be less than the discount rate for two reasons: mathematical and economic.

Mathematical reason: In DCF, the terminal value of the company is calculated by perpetually growing the expected annual cash flow at a fixed growth rate. If the growth rate is equal to or greater than the discount rate, this will lead to an infinite terminal value, which is mathematically impossible and logically unacceptable.

Economic reason: The discount rate typically represents the expected return rate of investors or the opportunity cost of investment. It reflects the risk borne by investors and the cost of giving up other investment opportunities. On the other hand, the growth rate is the expected future cash flow growth rate. Generally, the higher the risk of investment, the higher the expected return rate of the investor. Furthermore, the future growth rate of a company or investment usually slows down over time because the market becomes saturated and competition increases. High-speed growth is often difficult to sustain. Therefore, from an economic point of view, the discount rate is usually higher than the terminal growth rate.

In summary, in order for the DCF model to make sense mathematically and economically, the growth rate is typically set to be less than the discount rate. Therefore, we first calculate the discount rate r to understand the upper limit of the cash flow growth rate g, which will enable us to determine the cash flow growth rate more accurately.

It is also a difficult task to apply any discount rates according to CAPM to startup projects, as it is highly subjective to provide discount rates to any startup.

- Startups don't have historical data
- Highly unpredictable
- Variable cash flows and sometimes no cash flows

Lack of liquidity

Discount Rate r

We use the Capital Asset Pricing Model (CAPM), a theoretical model often used to quantify the risk-return of stock investments, to calculate the expected return of Ethereum, which is r. Let's briefly review CAPM.

CAPM is a single-factor model that only needs to consider the systematic risk of the market, which is the coefficient commonly known as Beta (β).

Beta is an indicator that measures the risk sensitivity of individual stocks relative to the entire market. This simplicity makes CAPM easy to understand and apply. According to CAPM, the expected return of a stock equals the risk-free rate plus Beta times the market risk premium. This relationship suggests that if a stock's Beta is high (i.e., the risk is high), then investors should expect a higher return to compensate for this risk.

CAPM provides a theoretical framework for investment decisions. For example, investors can use CAPM to calculate a stock's reasonable expected return and then compare this return with the actual market price to judge whether the stock is overvalued or undervalued.

However, it is also important to note that CAPM is a theoretical model, and its assumptions (such as all investors are rational, no transaction costs, etc.) cannot be fully realised in reality, so the model has certain limitations.

The formula for the CAPM model is as follows:

$$ERi = Rf + \beta\,(ERm - Rf)$$

Where:

E(R_i) is the expected return of asset i.

R_f is the risk-free return.

β_i is the beta coefficient of asset i, indicating the degree of association between the asset and the market.

E(R_m) is the expected return of the market.

Since we are comparing Ethereum to tech stocks here, we determine the parameters according to the following standards:

For R_f, the risk-free rate, we choose the U.S. 10-year Treasury bond yield, which is the same as most stock analyses. According to Ycharts data, as of September 27, 2023, the U.S. 10-year Treasury bond yield is 4.61%.

For E(R_m), the expected return of the market, we choose the 50-year average return of the S&P 500. Using the S&P 500's annual return provided by Macrotrends, we find that from 1973 to 2022, the average return is 8.69%.

The 50-year average return of the S&P 500:

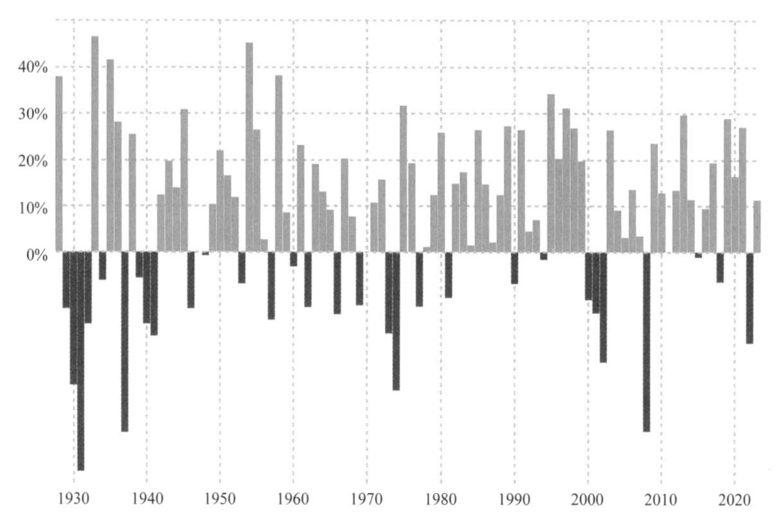

Chart: S&P 500 50-year returns.
Source: macrotrends

For Beta, we use the 300-week rolling data provided by Trading View (i.e., the Beta value as of September 28, 2023, is calculated based on the past 300 trading cycles), the Beta value is 1.6 (ETH Price relative to the S&P 500).

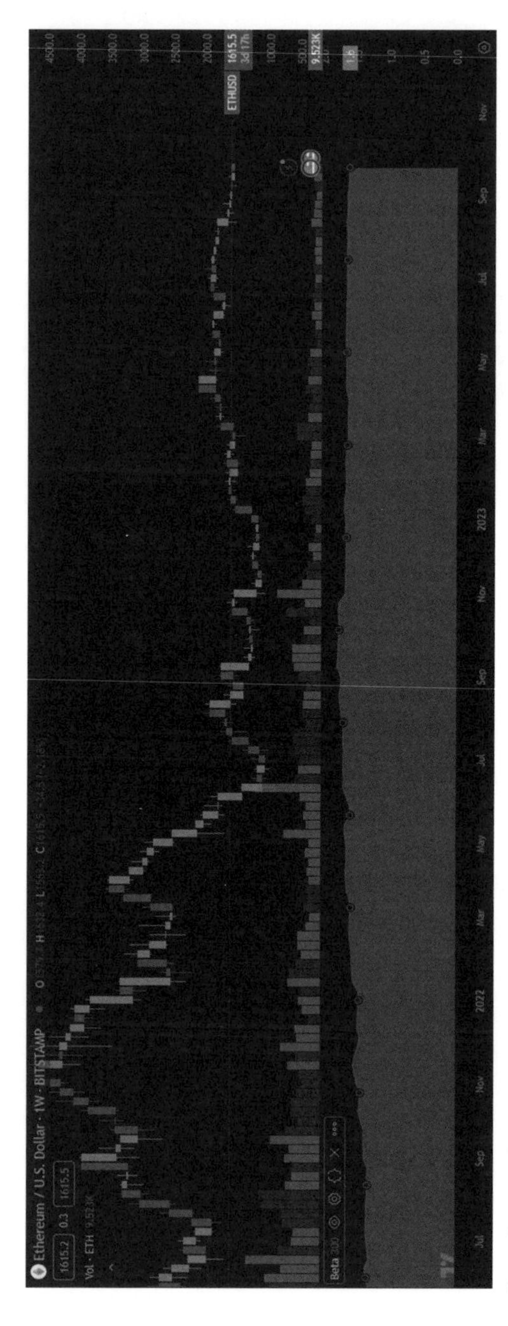

Chart: Ethereum Beta.
Source: Trading View.

With the above data, based on the formula

$$ERi = Rf + \beta (ERm - Rf)$$

We get

$$EReth = 4.61\% + 1.6 * (8.69\% - 4.61\%) = 11.14\%$$

The expected return of Ethereum is 11.14%.

According to the description above, the sustainable growth rate g is generally considered to be less than the discount rate r, so we determine that the sustainable growth rate of Ethereum's transaction fees is less than 11.14%.

Growth Rate g

Regarding the future growth of Ethereum's income, we believe that the important driving forces are the progress of Ethereum 2.0 and Layer 2.

To contextualise the future growth rate for Ethereum, we outline some of the phases of the Ethereum roadmap and their impact on adoptability.

As part of the Ethereum roadmap, there are a number of phases and upgrades that aim to improve Ethereum's scalability, robustness, decentralisation, and security. At the time of the writing of this book, Ethereum can process more than 1 million transactions per day, but the endgame is to be able to process over 1 billion transactions per day.

As part of the roadmap to achieve such scalability, Ethereum has made several upgrades, from which we highlight Olympic, Frontier, Homestead, Constantinople, Istambul, the beacon chain genesis, London, and the Merge. Posterior phases will continue to bring better scalability, decentralisation, and security to the network.

The benefits that Ethereum's endgame brings to Ethereum are primarily higher scalability and lower transaction costs. In the future, Ethereum sharding technology will break the network down into multiple small shards, each capable of independently processing transactions and smart contracts. This will greatly enhance Ethereum's processing power, allowing it to support more users and more complex applications. By improving network scalability, Ethereum will reduce transaction costs, making Ethereum more attractive to users and developers.

In addition to the progress of Ethereum itself, the application of Layer 2 solutions can significantly increase Ethereum's processing capabilities, reduce transaction costs, improve transaction speed, and make Ethereum more suitable for building complex dApps. Layer 2 solutions address Ethereum's main chain scalability issue by performing most calculations, storing them off-chain, and then submitting the results to the Ethereum main chain. Main Layer 2 solutions include state channels, side chains, Plasma, Rollups (both optimistic rollups and zk-rollups, etc). **An increase in transaction volume on Layer 2 directly increases the transaction volume on the Ethereum main chain and increases the number and volume of transactions relying**

on Ethereum for security. L2 chains also need to pay fees on the Ethereum chain in order to settle the data. Interactions between Layer 2 solutions and the Ethereum main chain will likely continue to increase main chain transaction volume, thereby affecting gas fees.

In fact, according to Vitalik's Endgame blog post, we can infer that in the future, L2 will process the majority of all use cases and transactions, and the Ethereum chain will be used mostly as the settlement and security layer for the L2 chains.

For example, Arbitrum is an Optimistic Rollup solution that provides Ethereum's Layer 2 expansion. It runs on top of the Ethereum main chain (Layer 1), processes most transactions, and periodically submits transaction results (also known as "state updates" or "rollback points") to the main chain. **Here are the main ways an L2 rollup interacts with the Ethereum main chain:**

Deposit: When users want to move their ETH or ERC-20 tokens from the Ethereum main chain to the Rollup, they need to execute a transaction on the main chain. This transaction usually involves sending a transaction to a smart contract with token data and how much assets the user wants to move to the Rollup. This process consumes Gas, so Gas fees need to be paid. Once the transaction is confirmed, the user's assets will be available on the Rollup.

Withdrawal: When users want to move their assets from the Rollup back to the Ethereum main chain, they need to start a withdrawal process on the Rollup. This process first generates a withdrawal record on the Rollup chain, and then this record is submitted to the Ethereum main chain. Once the withdrawal record is confirmed by the main chain, the user can execute a transaction on the main chain to actually receive their assets. This process also consumes Gas, so Gas fees need to be paid.

State updates: The Rollup regularly submits its state (or rollback points) to the Ethereum main chain. This is done by executing a special transaction on the main chain. The process of submitting the state requires gas.

Currently, the total locked value on Layer 2 has reached 10.7B USD, and it is expected to achieve more growth in the future, bringing income from Ethereum gas fees. At the time of writing this book, the top five projects on Layer 2 in terms of TVL are Arbitrum, OP, zkSync, Base, and dYdX, with the top two, Arbitrum and OP, accounting for nearly 80% of the entire Layer 2 market share and fees paid by scalability solutions.

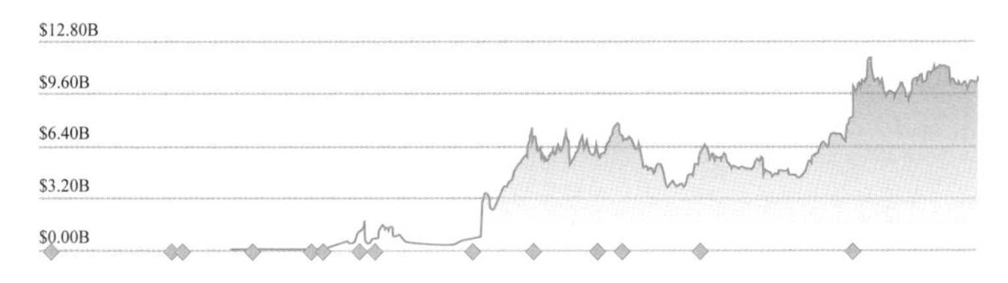

Source: L2Beat

#	NAME	RISKS	TECHNOLOGY	STAGE	PURPOSE	TOTAL	MKT SHARE
1	Arbitrum One		Optimistic Rollup	STAGE 1	Universal	$6.02B ▲ 6.36%	55.66%
2	OP Mainnet		Optimistic Rollup	STAGE 0	Universal	$2.70B ▲ 3.38%	25.02%
3	zkSync Era		ZK Rollup	STAGE 0	Universal	$465M ▲ 10.55%	4.29%
4	Base		Optimistic Rollup	STAGE 0	Universal	$462M ▼ 14.27%	4.27%
5	dYdX		ZK Rollup	STAGE 1	Exchange	$342M ▲ 0.93%	3.16%
6	Starknet		ZK Rollup	STAGE 0	Universal	$146M ▲ 8.92%	1.35%
7	Immutable X		Validium	n/a	NFT, Exchange	$96.80M ▼ 5.33%	0.89%
8	Mantle		Optimium	n/a	Universal	$90.70M ▲ 3.70%	0.84%
9	Loopring		ZK Rollup	STAGE 0	Tokens, NFTs, AMM	$88.85M ▲ 5.69%	0.82%
10	zkSync Lite		ZK Rollup	STAGE 1	Payments, Tokens	$71.10M ▼ 1.12%	0.66%

After the launch of Ethereum **Sharding** solutions, there is still a lot of room and opportunities for Layer 2 solutions. Layer 2 solutions can further enhance processing capacity based on Ethereum shards and can further reduce transaction costs by processing transactions off-chain. In addition, different Layer 2 solutions have different characteristics. For example, zkRollups provides a high degree of data availability and privacy, while State Channels are suitable for applications requiring high-frequency, low-latency transactions.

Using the DCF Valuation

Based on the previous conclusion, the perpetual growth rate g is less than the discount rate of 11.14%. We refer to the assumed growth rate proposed by Fidelity Digital Assets Research. For 2024-2025, the growth rate for the first two years is 25%; for 2026-2027, the growth rate is 20%, and it is assumed that the growth rate for

2028-2030 will be 10%. The perpetual growth rate after 2030 is assumed to be 5%. We can calculate the following cash flow and valuation of Ethereum's transaction fees. The Present Value – PV - obtained in this way can be regarded as the estimated market value of Ethereum in that year. Dividing by the supply of Ethereum can give the price of Ethereum.

Since Ethereum moved from Proof of Work to Proof of Stake in September 2022, new blocks are no longer created by miners by solving complex mathematical problems but are created by so-called validators. Validators need to pledge a certain amount of ETH (32 ETH at the time of writing the book) as an admission threshold for validation. Then, they are randomly selected to create new blocks and receive rewards for the job.

In addition, the implementation of EIP-1559 also had an impact on the supply of ETH. This proposal was integrated into the London hard fork in 2021 and introduced a new transaction fee structure in which a part of the transaction fee (base fee) will be burned, reducing the generation of new ETH.

Before the implementation of EIP-1559, Ethereum's transaction fees were determined by an auction mechanism where users bid, and the highest bid transaction was prioritised to be packaged into the block. This led to uncertainty in transaction fees. Users often have to predict the current network congestion and corresponding gas fees, which is quite difficult for ordinary users. EIP-1559 proposed a new fee structure where each block has a fixed base fee that automatically fluctuates according to network congestion. Users can also pay miners a "tip" to miners to prioritise their transactions. This new structure makes transaction fees more predictable. Users just need to look at the current base fee to know how much gas fee they need to pay. EIP-1559 also introduced a new mechanism where the base fee of each block would be burned instead of being paid to miners. This means that each transaction will reduce the total supply of ETH, making ETH potentially a deflationary asset. This feature is also an important reason why EIP-1559 has received much attention. The changes to Ethereum's fee structure due to EIP-1559 are shown in the following diagram:

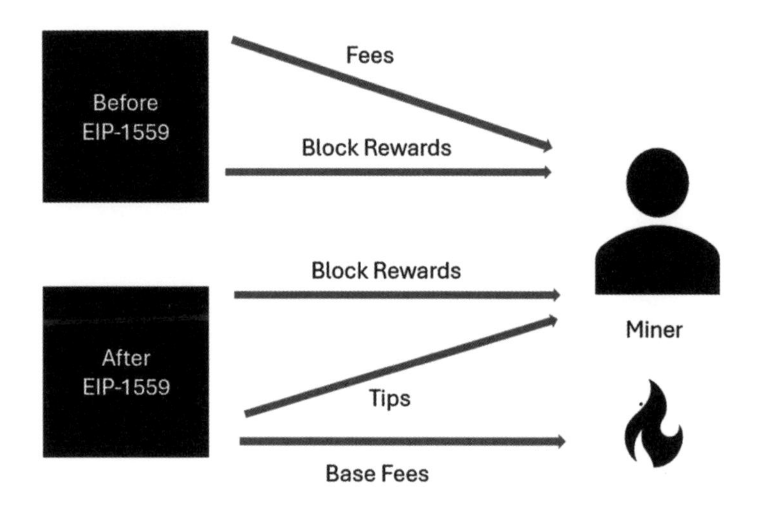

According to Etherscan data, the total supply of Ethereum on August 30, 2023, was 120,215,370.32, which fluctuated at 1% less compared to the supply a year before. In our valuation, we assume that the supply of Ethereum remains unchanged at 120,000,000.

Combining the data of Ethereum transaction fees from 2015 to August 30, 2023, the growth rate changes significantly and is quite affected by the overall market cycle. We refer to Fidelity Digital Assets Research and only focus on transaction data since the implementation of EIP1559, that is, after May 2021, take the average and then calculate the annual transaction fee for 2023. The total annual transaction fee for 2023 is 5,769,809,906 US dollars, which is used as the cash flow for the first period in 2023, and then the growth rate is superimposed.

Year	Total transaction fee (USD)	Growth rate
2015	686	
2016	159,273	23122%
2017	46,432,330	29053%
2018	160,210,604	245%
2019	34,716,862	-78%
2020	596,033,171	1617%
2021	9,914,346,459	1563%
2022	4,298,478,324	-57%
2023 (by 30 Aug 2023)	1,671,564,844	
Annualized 2023	5,769,809,906	34%

Data source: Etherscan, HashKey Capital arrangement

From here, we can get the market value and price of ETH for each year after 2023 as shown in the following table:

Year	Growth rate	Transaction Fee (Estimated)	Present Value	ETH Price (USD)
2023	-	5,769,809,906	181,777,900,085	1,514.82
2024	25%	7,212,262,383	188,377,687,690	1,569.81
2025	25%	9,015,327,978	210,909,043,580	1,757.58
2026	20%	10,818,393,574	225,388,983,057	1,878.24
2027	20%	12,982,072,289	239,678,922,195	1,997.32
2028	10%	14,280,279,518	253,397,081,839	2,111.64
2029	10%	15,708,307,470	267,345,237,238	2,227.88
2030	10%	17,279,138,217	281,419,189,196	2,345.16
After 2030	5%	295,490,148,656	295,490,148,656	2,462.42

As illustrated in the table above, the Perpetual Value for Ethereum after 2030 would return an ETH price of $2,462.

Keeping the discount rate at 11.14%, small changes in the perpetual growth rate result in significant differences in the derived prices:

Terminal Growth Rate	10%	8%	6%
Transaction Fee (Estimated) after 2030	1,515,713,878,654	550,291,026,008	336,170,004,215
ETH Supply	120,000,000	120,000,000	120,000,000
ETH Price after 2030 (USD)	12,630.95	4,585.76	2,801.42
Terminal Growth Rate	4%	2%	
Transaction Fee (Estimated) after 2030	242,004,736,928	189,049,652,261	
ETH Supply	120,000,000	120,000,000	
ETH Price after 2030 (USD)	2,016.71	1,575.41	

The table above shows that different assumptions about the terminal growth rate have a significant impact on the price of ETH. The price corresponding to a 10% terminal growth rate is nearly 10 times the price corresponding to a 2% terminal growth rate.

DCF can be a very useful method for the valuation of revenue-generating chains. Although the DCF method is typically used for the valuation of traditional businesses, Ethereum (as well as other blockchain protocols) can, in certain aspects, be seen as a traditional business.

This analysis employed the Discounted Cash Flow (DCF) model for valuation, commonly used in traditional stock analysis, and the Capital Asset Pricing Model (CAPM) to calculate the expected return of Ethereum. The DCF and CAPM models provide a structured framework to approach Ethereum valuation, even though the application of these models to a crypto asset is innovative and not without challenges, especially considering the high unpredictability and variable cash flows of such assets.

Limitations

It's worth noting that there are some limitations to using cash flow forecasts. The greatest limitation is the uncertainty of the forecast, where the assumptions of growth rates and the choice of discount rates are often subjective.

From here, we can also infer that discount rate variations can have a high impact on the derived prices. Consequently, a reduction in future risk-free rates (here as treasury bill rate) will positively impact the model.

A savvy investor will consequently always take into consideration the risk-free rates whenever applying this model.

The future expansion of ETH, continuous upgrades, and exploration of MEV resistance will all impact its transaction fees and thus affect its growth rate. In terms of the discount rate, we regard Ethereum as a type of tech stock, which is a subjective assumption, and we have also mentioned the limitations of CAPM itself. Furthermore, DCF valuation primarily focuses on the intrinsic value of Ethereum and overlooks the information that the market may reflect, such as crypto industry trends and the macroeconomic environment. These are all very easy factors affecting ETH transactions and prices, and DCF analysis may not fully capture market conditions.

Limitations of applying DCF to low transaction fee chains

According to Solana Beach data, Solana's current TPS is 5273, the total transactions processed are close to 240 billion, and the total staked exceeds 400 million (locked tokens can also be staked).

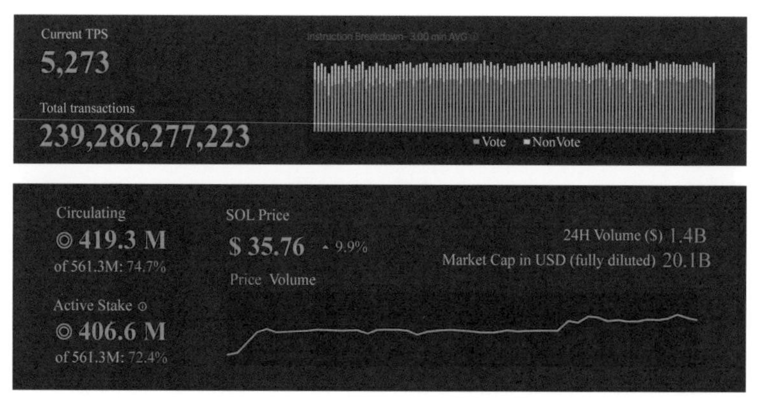

Data source: Solana Beach

Regarding the ecosystem, according to DeFiLlama data, Solana's TVL accounts for 0.97%, the highest TVL among non-EVM chains, accounting for 17.59%, and there are hundreds of ecosystem projects, including DeFi, NFT, socialfi, gaming, wallets, and other sectors, and further subdivided into lending, Dex, derivatives, synthetic assets, options, stablecoins, among others.

Name	Category	TVL	1d Change	7d Change	1m Change
1 Marinade Finance 1 chain		$313.23m	+2.99%	+20.83%	+132%
2 Jito 1 chain	Liquid Staking	$184.08m	+8.03%	+99.47%	+216%
3 Solend 1 chain	Lending	$74.19m	+3.33%	+11.02%	+41.93%
4 Lido 5 chains	Liquid Staking	$72.32m	+2.67%	+13.51%	+18.64%
5 marginfi 1 chain		$52.92m	+2.49%	+30.64%	+121%
6 Orca 1 chain	Dexes	$52.28m	-3.89%	+4.31%	+20.44%
7 Raydium 1 chain	Dexes	$35.28m	+0.12%	+5.85%	+7.13%
8 SPL Governance 1 chain	Services	$32.42m	+0.03%	-0.47%	+0.29%
9 Drift 1 chain	Derivatives	$23.83m	-1.83%	+8.09%	+22.46%
10 JPool 1 chain	Liquid Staking	$23.07m	+3.34%	+71.57%	+118%

Data source: DeFiLlama

DCF model applied to Solana

We follow the previous Ethereum's thinking, consider Solana as an asset analogous to technology stocks, and still choose the 10-year US Treasury bond yield of 4.61% for risk-free returns. E(R_m), and the expected return of the market, we continue to use the 50-year average return of the S&P 500 of 8.69%. According to the formula $\beta = Cov(Ra, Rm)/Var(Rm)$, we calculate the relationship between the S&P 500 and Solana and get a β value of 0.5.

We get

$$ER_{sol} = 4.61\% + 0.5 * (8.69\% - 4.61\%) = 6.65\%$$

The perpetual growth rate is less than 6.65%, and the perpetual growth rate should be less than the discount rate. So, we set the perpetual growth rate at 5%. We estimate that the growth rate of the previous stages is the same as ETH. As for the supply part, Solana's maximum supply is 560.21M, according to its supply release mechanism. The initial inflation rate is 8% and then decreases by 15% each year. From this, the long-term inflation rate is 1.5%.

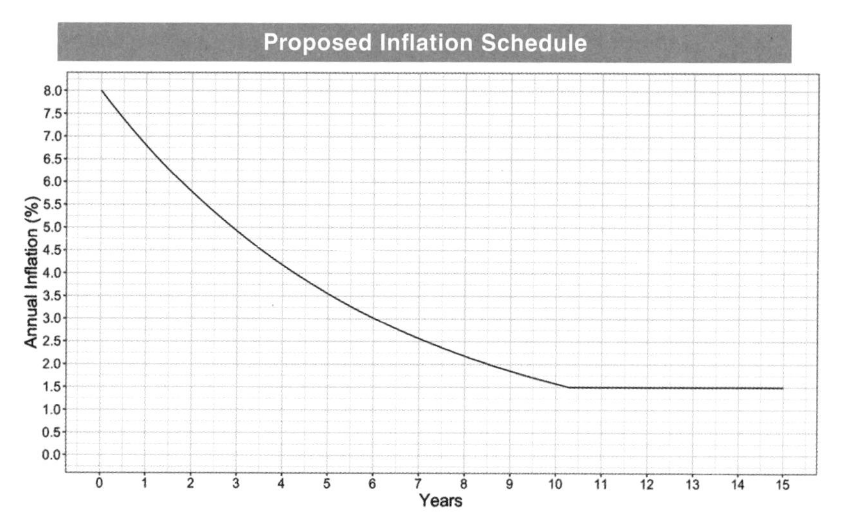

Source: solana.com

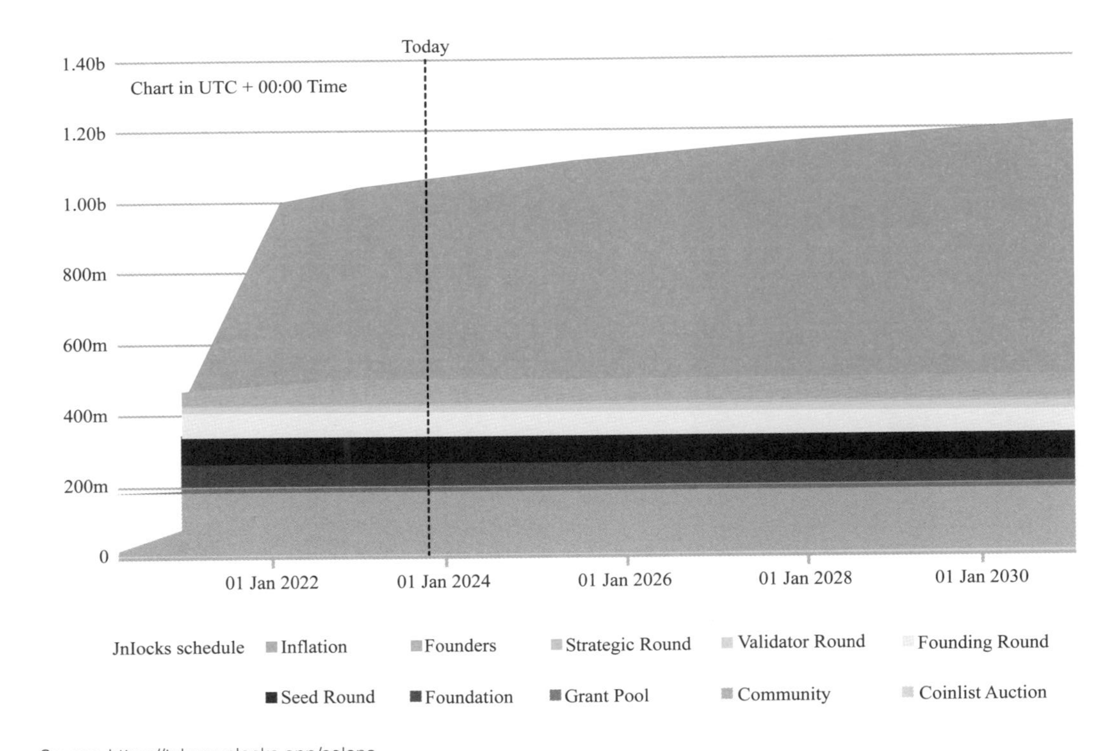

Source: https://token.unlocks.app/solana

Regarding Solana's cash flow, we use the data from the token terminal and conclude that the annual transaction fee income in 2023 is about 14.88 million USD.

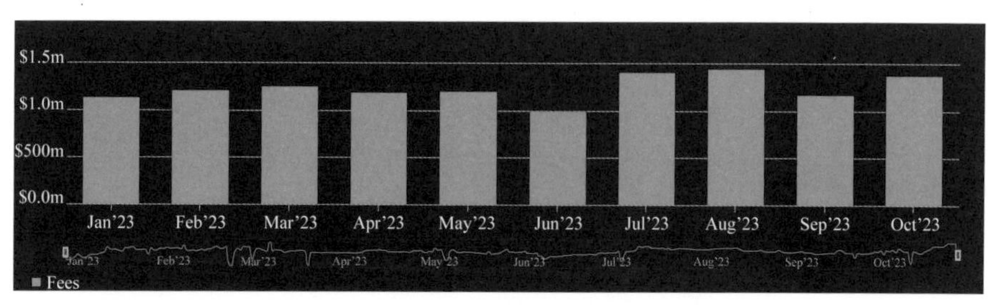

Data source: token terminal

From this, we get Solana's future cash flow and **price estimation**:

Year	Growth rate	Transaction Fee (Estimated)	PV
2023	-	14,886,557	1,867,857,051
2024	25%	18,608,197	1,954,173,180
2025	25%	23,260,246	2,090,057,459
2026	20%	27,912,295	2,205,786,034
2027	20%	33,494,754	2,324,558,510
2028	10%	36,844,230	2,445,646,897
2029	10%	40,528,653	2,571,438,186
2030	10%	44,581,518	2,701,910,172
After 2030	5%	2,837,005,681	2,837,005,681

Year	Circulating Supply	SOL Price (USD)	
2023	1,069,736,988	1.74608999	
2024	1,098,919,989	1.77826703	
2025	1,124,064,775	1.85937457	
2026	1,146,812,764	1.92340555	
2027	1,166,398,424	1.99293694	
2028	1,188,593,832	2.05759683	
2029	1,199,234,698	2.14423264	
2030	1,212,240,518	2.22885651	
After 2030			

It can be seen that, unlike Ethereum, as mentioned earlier, Solana's transaction fees are extremely low, so the predictions for cash flow and price are relatively low. The DCF model only has a reference value for Solana's valuation.

The prices above are notably far from the current price, sitting at $60 at the time of the writing. This illustrates the limitations of the DCF model in certain chains, depending on their pricing and business model.

Comparing Ethereum with the Cryptocurrency Market

In the previous discussion, we introduced the method of estimating the cash flows of Ethereum by valuing Ethereum as a tech stock. Now, we will compare Ethereum to the entire crypto market. Please note that the selection criteria for the risk-free rate and expected market return in the cash flow estimation may vary based on different methods. Consequently, the expected return of Ethereum will also experience significant variations.

Once again, we confirm the following parameters based on the CAPM formula:

$$ERi = Rf + \beta (ERm - Rf)$$

R_f risk-free rate: we choose the five-year return of the top 5 tokens by market value (excluding stablecoins). According to the Cryptocurrency Top 5 Equal Weight Index launched by S&P Dow Jones Indices, the five-year return is 4.37%.

E(R_m) Expected market return: We chose the S&P Cryptocurrency Broad Digital Market Ex-MegaCap Index launched by S&P. The three-year return is 5.09%. S&P explains the index as including tokens and cryptocurrencies with a market capitalisation value exceeding $80M, having more than three months of trading time (tracked by data provider Lukka), and meeting the median daily trading volume of $100,000 for three months. Both liquidity and market value need to meet the requirements.

Excluding Bitcoin and Ethereum, the two largest tokens by market value, other requirements include having a white paper and not being a stablecoin or other pegged asset. Since most tokens other than Bitcoin and Ethereum haven't been around long, we use the three-year return data as a reference.

To calculate β, according to the formula β = Cov(Ra, Rm)/ Var(Rm), calculating the relationship between the S&P Ethereum Index and the Cryptocurrency Broad Digital Market Ex-MegaCap Index, the β value is 0.11.

Under this method, the discount rate r for ETH is calculated to be 4.45%.

$$ER_{eth} = 4.37\% + 0.11 * (5.09\% - 4.37\%) = 4.45\%$$

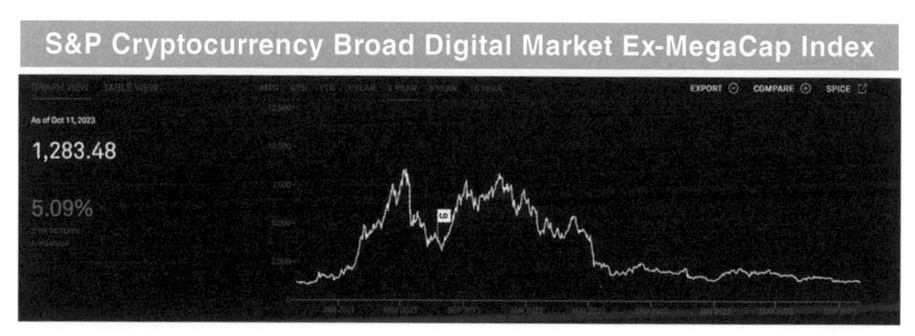

Source: spglobal.com

Assuming the growth rate assumptions for 2030 and the previous stages remain consistent, the perpetual growth rate needs to be lower than the discount rate of 4.45%, we assume it to be 4%, and we can get the market value and price of ETH for each year after 2023 as shown in the following table:

Year	Growth rate	Transaction Fee (Estimated)	PV	ETH Price (USD))
2023	-	5,769,809,906	2,785,891,935,288	23,215.77
2024	25%	7,212,262,383	2,894,600,574,090	24,121.67
2025	25%	9,015,327,978	3,026,114,251,204	25,217.62
2026	20%	10,818,393,574	3,151,761,007,404	26,264.68
2027	20%	12,982,072,289	3,281,195,978,659	27,343.30
2028	10%	14,280,279,518	3,414,227,127,421	28,451.89
2029	10%	15,708,307,470	3,551,879,955,073	29,599.00
2030	10%	17,279,138,217	3,694,230,305,604	30,785.25
After 2030	4%	3,841,344,415,987	3,841,344,415,987	32,011.20

With the discount rate held at 4.45%, slight changes in the perpetual growth rate result in price differences:

Terminal Growth Rate	3%	2%	1%
Transaction Fee (Estimated) after 2030	1,227,414,645,735	719,376,366,571	505,853,031,850
ETH Supply	120,000,000	120,000,000	120,000,000
ETH Price after 2030	10,228.46	5,994.80	4,215.44

Similar to the previous point we mentioned, the assumption of the terminal growth rate has a significant impact on the estimation of ETH price. The price corresponding to a 3% terminal growth rate is more than twice the price corresponding to a 1% terminal growth rate.

2.3 Calculating the price equilibrium for Ethereum

As in any other market/asset, two major forces influence the price of Ethereum: the supply and the demand. However, unlike any other assets, Ethereum's supply and demand are inversely correlated and have a recursive effect. This recursive effect on Ethereum shows that the higher the demand, the lower the supply (in fact, the higher the demand is, the higher the deflation rate). In other words, it means that, unlike most assets, the higher the ETH price, the lower the quantity supplied.

To make even more clear what happens to the supplied quantity of Ethereum, let's take a look at some examples of different types of supply.

Elastic supply: If the demand for a certain car model increases, putting pressure on the price, car manufacturers will ramp up car production to increase the number of cars in the market and satisfy the demand. This means that the car industry has an elastic supply. The supply is elastic because when demand and prices increase, car manufacturers will simply produce more of it. There are no limitations on how many cars can be produced other than market demand and raw materials. This is also the reason why, in most cases, a car is a depreciating asset.

Inelastic supply: If the demand and the price of gold increase in a given year, no matter how much it increases, gold production is always very limited. The world's gold production rate is, on average, 2%/year and doesn't deviate much from it. Of course, when the gold price increases a lot, gold mining becomes more profitable, which pushes annual gold production slightly up to close to 2.5%. In the case of gold, we say that gold's supply is mostly inelastic because no matter how high the demand/price goes, the gold supply increase is always very limited to the mining capacity, which averages 2%/year.

Shrinking supply: If the demand (and price) of a cryptocurrency such as Ethereum increases, the more the demand increases, the more the supply will contract. In fact, Ethereum supply becomes deflationary when there's more activity on the network, which usually happens in times of higher demand and high prices. This translates into Ethereum supply being inversely correlated with the price. Almost no assets in the world behave like this. Unlike any other asset, the higher the Ethereum price goes, the scarcer the asset becomes.

Why does Ethereum supply behave like this? Since the EIP-1559, part of the fees are burned. Meanwhile, the ETH emissions were lowered after the Merge (shift from PoW to PoS), and the fee burn rate has been, for most of the time, higher than the emissions, meaning that the inflation rate is negative. In fact, Ethereum enters a deflationary state whenever the gas fees are over 20 Gwei in the Ethereum blockspace market. In other words, whenever the fees are over 20 Gwei, which is lower than the average fee, the burning mechanism burns more ETH than what's issued, and the asset becomes deflationary.

Saying this, although Bitcoin's supply is perfectly inelastic, Ethereum's supply inelasticity goes one step further by increasing its deflation when the demand increases.

This can create a positive feedback loop for the price, as the price can become even more sensitive to demand increases, and demand surges will be tied to a shrinking supply.

In addition, as the total circulating supply decreases, fees burned have a higher deflationary impact on the supply. We can effectively say that Ethereum deflation will have a compounding effect.

Here's how the deflationary compounding effect works: Let's say if we have ten units of a product and burn 1 unit; the inflation rate is -10 %. However, the next unit being burned will have a higher percentage impact on the supply. Now, we have nine units, and burning 1 unit will represent an inflation rate of -11.1%.

This incremental deflation rate can already be verified on Ethereum. Although its impact is still small, it will compound over time.

For example, Ethereum's circulating supply was 120,522,097 ETH in December 2022. Burning, on average, 87,000 ETH monthly would represent an annual inflation of -0.8662%. However, in July 2023, the supply dropped to 120,217,769, and now, the same 87,000 ETH burn rate represents -0.8684%. The difference between -0.8662% and -0.8684% may not seem much, but future compounding effects can be powerful. By 2030, the yearly inflation will be -0.93%.

This deflationary compounding effect will, over the long term, have a positive impact on the price.

Calculating price equilibrium according to the supply and demand, considering the staked % and the Ethereum deflation caused by fee burning:

Supply calculation

To calculate the circulating supply, we add/subtract the inflation rate from the supply of the previous period, as per the equation below.

$$S = P + (K*N)$$

Where

S = Circulating supply
P = supply previous period

K = represents the inflation (or deflation) rate for the period. We find that (-1%*8 700 000) offers the most realistic rate according to historical data, which represents a deflation of approximately -87 000 ETH per month. Through a linear regression between supply and market capitalisation, we found that -87,000 is the average monthly burn rate.

N = the number of periods in our model expressed in months.

It makes sense that k = 8 700 000 is a constant because Ethereum's blockspace is somehow constant, and, on average, Ethereum users will pay the same amount of ETH fees for blockspace, which generates a similar amount of fees burned.

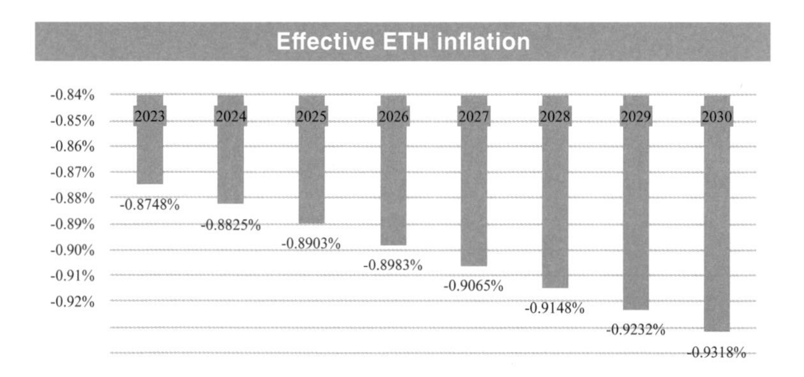

Considering the historical inflation rate and circulating supply calculation, the deflation rate would increase as the circulating supply decreases, and k is a constant. In fact, as we saw before, it is expected that through the years, the deflation rate will accelerate.

Demand calculation:

Calculating the demand is more demanding than calculating the supply.

We need to start by calculating an expected rate of return:

$$ERi = Rf + \beta i * (Rm - Rf)$$

Where:

Eri = Expected rate of return
Rf = risk-free rate

βi (beta of i) = risk of the investment vs the overall market. A beta less than 1 indicates the investment will be less volatile than the market, while a beta greater than 1 indicates the investment's price will be more volatile than the market.

In this simulation, we will use a Beta of 1.6, which is the beta of Ethereum vs S&P 500 over the last 7 years.

(Rm - Rf) is the Market Risk Premium. This is the return expected from the market above the risk-free rate.

In this model, we will use two different rates as the Rf risk-free rate:

- US 10-year T-bond. Rate 4.25% as of Sept. 2023
- *Aave USDC supply rate. Rate 3.1% as of Sept. 2023*

Then we calculate the Beta of i:

```
Beta of i: Beta = Corr((Close - Close[1])/Close, (Ovr - Ovr[1])/
                  Ovr) * (Stdev((Close - Close[1])/
                  Close, Length)/Stdev((Ovr - Ovr[1])/
                  Ovr, Length))
```

Where:

Close is the current closing price of the security (for which the Beta is being calculated).

Ovr is the current closing price of the market index (SPY).

Close[1] and Ovr[1] are the previous closing prices of the security and the market index, respectively.

Stdev is the standard deviation function.

Corr is the correlation function.

Length is the time window for the calculation.

Risk market premium - This is the return expected from the market above the risk-free rate.

Market Risk Premium = Market Return - Risk-free rate

To calculate the market return, we estimate that the Ethereum network will capture a percentage of the global market for finance, gaming, and bond markets.

2030 market capture		Global size
Finance	1%	$115,000,000,000,000
Metaverse/Gaming	2%	$145,000,000,000
Bonds	1%	$119,000,000,000,000
Total capture 2030		$2,342,900,000,000
Sept-2023 mcap		$196,606,475,789
Total target 2030 mcap		**$2,539,506,475,789**

Here is an example of a possible market captured by the Ethereum network.

This step is important so that we cap the total future return. According to the total target future market capitalisation, we can now set a Risk Premium that will match the demand growth to reach the future market capitalisation.

Demand = D = Mcap = Mcapl * (1+ ERi*(0.618*s)^-0.0618*c)-(c-1)

Where,

Mcap1 = market capitalisation of the previous period

ERi= monthly growth rate, which is the annual growth rate divided by 12

c = a constant that represents the slope of the diminishing demand growth curve. We have set n = 3.33

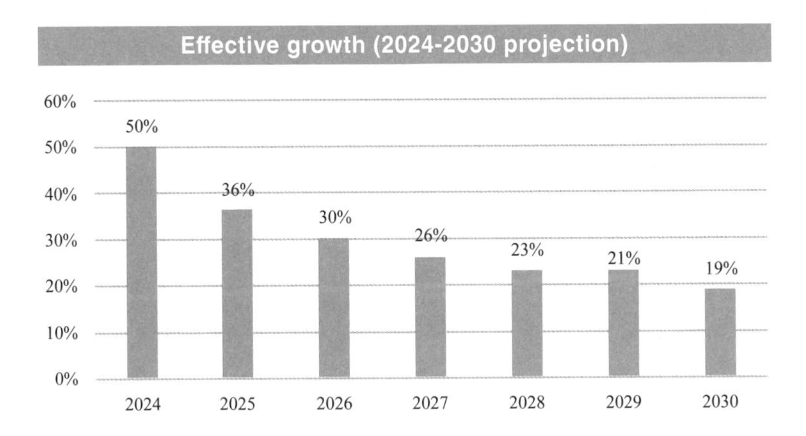

The second part of the above demand function is a rational function that illustrates diminishing growth over time. This diminishing growth is more realistic than linear growth. Let's say we believe that Ethereum demand will grow by 50% in 2024. Most likely, the growth will slow down as the network will have less room to grow as it expands.

Note that this is the growth that should be required to reach the total target 2030 market capitalisation that we saw in the table on the previous page.

Price calculation:

The price calculation follows a simple supply and demand equation.

$$P = qS/qD$$

The price is simply the result of the supply that we have calculated before (in this case, the circulating supply of ETH) divided by the demand (that we just calculated).

Practical example

These are possible projections for the ETH price according to the above model. The projections are for April 2030.

Assumptions:

Expected rate of return		
ERi = Rf + βi*(Rm - Rf)		
Rf risk free rate	4%	
βi Beta	1.70	
Rm risk premium	38%	
Eri	62%	
2030 market capture		Global size
Finance	1%	$115,000,000,000,000
Metaverse/Gaming	2%	$145,000,000,000
Bonds	1%	$119,000,000,000,000
Total capture 2030		$2,342,900,000,000
Sept-2023 mcap		$196,606,475,789
Total target 2030 mcap		$2,539,506,475,789

According to our assumptions and model, we would have the following ETH price by April 2030:

Apr/2030 Diminishing/increasing Growth Projection	
ETH Circ. supply	108,907,769.89
Price	$12,321
Mcap	$2,560,834,427,005

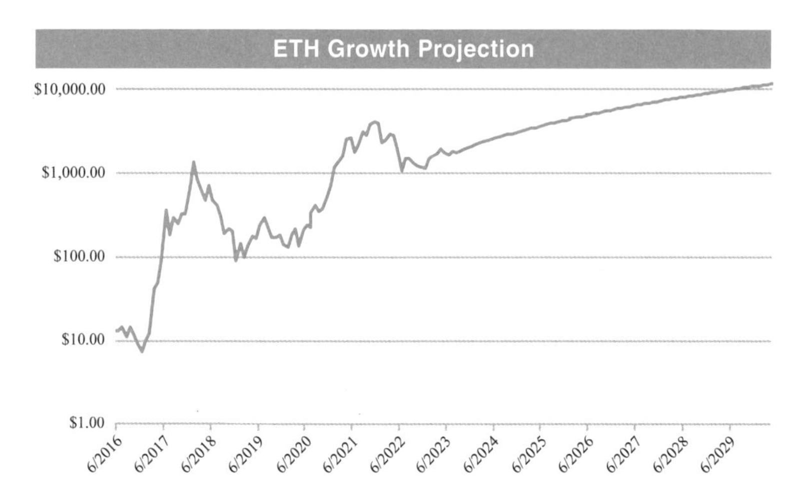

The current model implies a price of $12,321 by April 2030. This assumes a 4% risk-free rate, 1.6 Beta, close to -1% inflation rate, and a market cap reaching $2.5 trillion, which would translate into a risk-free premium of 40% and an Eri of 62% to start.

This projection is close to the VanEck made in May 2023, which puts a $11,800 price target for Ethereum in 2030.

Conclusion

Ethereum's unique supply dynamics, characterised by an inelastic supply with deflationary pressures, positions it for a potential increase in value over time. As the demand for Ethereum grows, driven by its applications in finance, gaming, and other markets, the reduction in supply from fee burns becomes increasingly impactful. This deflationary compounding effect is not just a theoretical concept but is already observable in the incremental changes in Ethereum's inflation rate.

The detailed model presented here aims to provide a framework for understanding how Ethereum's price could evolve, taking into account both supply and demand factors. By employing a combination of historical data, regression analysis, and market capture scenarios, we arrive at a projection that suggests a significant increase in Ethereum's price by April 2030. While these models are based on assumptions that must be rigorously tested and revised as market conditions change, they offer a compelling case for the long-term value proposition of Ethereum.

It is imperative, however, for investors and stakeholders to continuously monitor the variables at play, including technological advancements, regulatory changes, and shifts in market sentiment, which could all influence Ethereum's trajectory. The interplay between deflationary supply and escalating demand, underpinned by the robust utility of the Ethereum network, paints a bullish picture for the future, but one that is not without its risks and uncertainties. As with any predictive model, the outcomes here should be considered as one of many possible futures, each contingent upon a multitude of evolving factors.

2.4 Value of a perpetual bond

In traditional finance, a perpetual bond is a bond with no maturity date. The issuer pays coupons to the bondholders indefinitely.

To adapt the concept of a perpetual bond to Ethereum, we would need to think about what the "coupon" or regular payment would be and what the "principal" or initial investment is.

For the "coupon," we can consider Ethereum staking rewards. In Ethereum, holders can "stake" their ETH to help secure the network and earn rewards. This is somewhat similar to the coupon payment of a bond.

The "principal" is the initial amount of ETH that is at stake. However, unlike a bond, the investors can get the principal back when they stop staking, assuming the network is not slashed (a rare event where stakers are punished for malicious behaviour on the network).

Since Ethereum became a proof of stake network, the community often calls it "the internet bond."

Investors can see Ethereum as a bond. To be more precise, we should say, "Investors can see staked Ethereum as a bond." However, considering that it is extremely easy to skake ETH and contribute to the network security, thus receiving the respective APR, we assume that is variable is taken for granted. Additionally, it makes sense that all rational actors stake ETH in order to get the additional yield, as the risks of doing so are extremely low.

Ethereum indeed pays a coupon (Ethereum staking rewards), and it has no maturity. Consequently, we can perhaps evaluate it the same way as value perpetual bonds.

For the sake of calculating the "Annual Coupon Payment" – i.e., the rewards paid to ETH stakers, we will ETH issuance, tips/ fees, and MEV.

There's one important note before moving forward: the Ethereum network is more than a simple bond. It's a smart contract chain that provides distributed computing to thousands of dApps. Considering this, calculating the value of a perpetual bond somehow reduces the true value of ETH (which, as a result, should be higher than the value of a perpetual bond).

The formula for the price of a perpetual bond is C/r, where

C is the annual coupon payment (annual staking rewards in the case of Ethereum)

r is the required return on investment or discount rate

So, if an investor stakes P ETH and expect to get a return of R ETH per year from staking rewards, and expects a return of r per year on your investment, the value of your staked ETH as a "perpetual bond" would be R/r.

Value of Perpetual Bond = Annual Coupon Payment/
Discount Rate

Where Discount Rate = Risk-Free Rate + Beta *
(Market Return - Risk-Free Rate)

Simple calculation:

To calculate the value of Ethereum as a perpetual bond according to the formula above, we assume that:

Current ETH price: $2100

Coupon 5% = $105 = $2100 * 5%

Discount rate 4%. We use the 10-year treasury rate, considering that the value of a perpetual bond should be seen from a very long-term perspective.

$105/4% = $2625

According to the simple calculation of Ethereum as a perpetual bond, the ETH price should have been, at the time of writing this book, $2625.

We can also add additional variables to the calculation of Ethereum as a perpetual bond. For the following calculation, we will derive the risk premium from using the CAPM – Capital Asset Pricing Model.

The complete calculation, which includes a complete discount rate calculation, looks like this:

Coupon 5% = $105 = $2100 * 5%

Discount rate 4% (the 10-year treasury rate)

Risk-free rate: 4%

ETH Beta: 1.6 (compared to the S&P 500)

Market return (S&P 500): 9%

= $105/4% + 1.6*(10%-4%)

= $875

According to this more complete version where we include the CAPM calculation, the value of ETH as a perpetual bond at the time of this writing should be $875.

Another important consideration when using the "value of a perpetual bond" to evaluate Ethereum is that the higher the staking ratios, the lower the yields. APY is approximately inversely proportional to the square root of ETH staked.

We can calculate the approximate staking rewards (excluding fees and MEV) with

$$APR = 143/\text{sqrt}\ (\text{total ETH staked})$$

According to this, and considering the average fees paid on the Ethereum network, if the totality of ETH circulating supply is staked, the staking yields would be close to 1.5%. If we look exclusively at the "value of a perpetual bond," in this scenario, ETH would be worth less (around $250, according to the model).

However, what the model misses is that more ETH staked means less ETH circulating, which equals lower supply, which translates into higher prices.

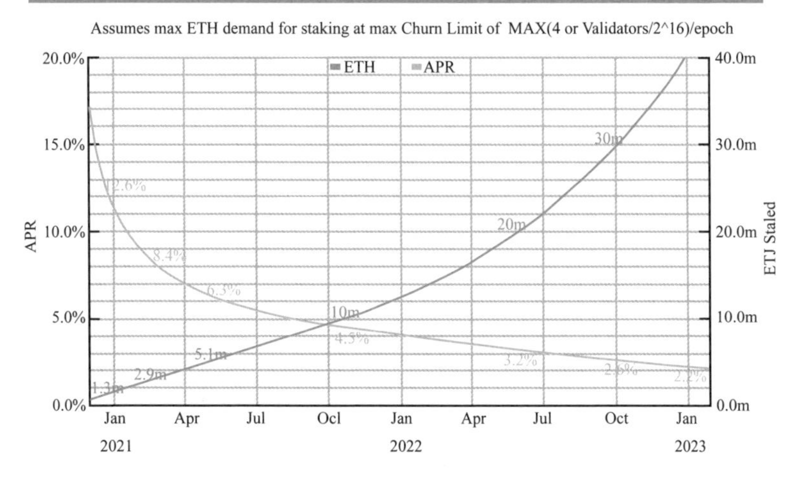

Value of Perpetual Bond = Annual Coupon Payment/ Discount Rate

Where Discount Rate = Risk-Free Rate + Beta * (Market Return - Risk-Free Rate)

The value of staking in the long term should be:

Value perpetual = (((143/sqrt (total ETH staked)) * total staked)/discount rate)/total staked

= (((143/sqrt (29 000 000) * 29 000 000) / (6%+1.6*(10%-6%))/29 000 000

= 0.28

In other words, according to this model, the value of staking is 0.28 ETH. From here, we can calculate the present value:

Current price + staking value = value of the Ethereum bond

Conclusion

The efficient market hypothesis suggests that financial markets are "informationally efficient," which means that it's impossible to consistently achieve greater than average market returns. By suggesting that Ethereum is undervalued based on its staking returns, the analysis appears to contradict this widely accepted theory. However, it is sensible to attribute a value to the future yields provided by staking Ethereum. Additionally, at the time of writing this book, most market participants still don't see Ethereum as a bond, thus it still might be possible to take advantage of the market misalignment.

2.5 Metcalfe's Law:

What is Metcalfe's Law?

Metcalfe's Law, formulated by Robert Metcalfe in 1980, states that the value or impact of a telecommunications network is proportional to the square of the number of connected users (n^2). Initially, the law referred to "compatible communicating devices" such as fax machines and telephones, but it later became associated with Ethernet users after an article by George Gilder was published on 13 September 1993 in Forbes.

Metcalfe's Law posits that the value of a network is proportional to the square of its number of users. Initially used to describe the value of communication networks, this theory has also been applied to the valuation of social networks and blockchain networks. It is represented by the formula **V = n^2*a**, where V is the value of the network, n is the number of users on the network, and a is a coefficient (Since the square of the number of users cannot be directly compared with the market value of Ethereum).

Metcalfe's Law uses the square of the number of users to represent the value of a network because each new user can establish new connections with all existing users in the network. As such, the potential number of connections in a network increases exponentially with the growth of user numbers. Specifically, if a network has n users, each user can establish a connection with the other n-1 users. Therefore, the total number of potential connections in the network is n*(n-1). However, since each connection involves two users, the actual number of connections should be n*(n-1)/2. This formula can be approximated to n^2 (when n is large enough).

Therefore, Metcalfe's Law uses the square of the number of users to represent the value of a network, reflecting that each new user can significantly increase the potential number of connections in the network, thereby increasing the value of the network.

Valuing Ethereum Using Metcalfe's Law

For a blockchain network like Ethereum, an increase in the number of users can bring about a larger network effect. Each new user may bring new transactions, new smart contracts, and new DApps, all of which increase the activity on the Ethereum network and, therefore, enhance its value. The value of a DApp supported by Ethereum often depends on its number of users; a DApp with more users will have greater value, thereby enhancing the overall value of the Ethereum network.

The chart below, created using Glassnode, shows a metric of the number of users(square) and the market capitalisation of Ethereum. We can roughly see that before January 2021, changes in the market value of Ethereum generally kept pace with fluctuations in the number of active addresses (squared). However, after January 2021, the two showed significant divergence. Considering that more recent times have greater statistical significance for future valuation, we choose to analyse the data after January 2017.

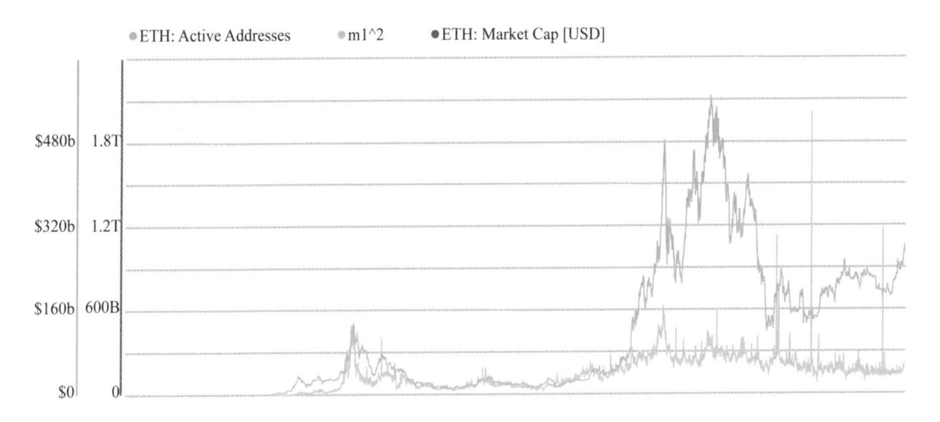

Data source: Glassnode, HashKey Capital collation

Our approach to valuing Ethereum using Metcalfe's Law involves determining the number of users (square), the market value of Ethereum, and the constant 'a'. We choose the number of **active Ethereum addresses** as the number of users. The constant 'a' is calculated by dividing the price by the square of the daily active user count, using a 1-year **average** value from

Nov. 2022 until Nov 2023. The constant 'a' is calculated as **0.0000000101361.**

a = v/n^2
a = [1-year average price]/
[1-year average number of active wallets]^2
a = $1672.78/406236^2
a = 0.0000000101361

Using the 1-year average allows us to remove the noise from daily variations of the number of wallets and price.

We can then evaluate the asset using the formula **V= a*n^2.**

A regression analysis comparing the Metcalfes valuation with the market prices gives us an R^2 of approximately 0.68, indicating a certain degree of correlation.

However, because both the actual market value and the estimated market value fluctuate on a daily basis, when we apply a **moving average of 60 days** to both sets of data, the R^2 value reaches 0.86. Therefore, there is a strong correlation between V= 0.0000000101361*n^2 and the actual market value on a given day. Hence, we use a = 0.0000000101361 to estimate the future market value of Ethereum.

For example:

V = a*n^2
V = 0.0000000101361*400,000^2
V = $1,621

In November 2023, the number of active addresses is about 400,000 and in mid-January 2024 was close to 500,000. We estimated 1 million, 2 million, 5 million, 10 million, and 20 million.

At the time of the writing, the number of active addresses has been fluctuating between 400,000 and 500,000 and the ETH price is between $2,200 and $2,500, demonstrating that the model can be close to the reality.

Number of address	Market cap	Price USD
200,000	$48,653,191,283	$405.44
300,000	$109,469,680,386	$912.25
400,000	$194,612,765,131	$1,621.77
500,000	$304,082,445,517	$2,534.02
1,000,000	$1,216,329,782,068	$10,136.08
2,000,000	$4,865,319,128,273	$40,544.33
5,000,000	$30,408,244,551,707	$253,402.04
10,000,000	$121,632,978,206,827	$1,013,608.15
20,000,000	$486,531,912,827,307	$4,054,432.61

Limitations

Although Metcalfe's Law provides a method for quantifying network value, it is merely a simplified model and has some limitations, particularly when applied to complex network systems.

Not all users will interact with all other users in the network, and the contributions of different users to the network are not equal. In the assumptions of Metcalfe's Law, the contributions of users are equal. However, in reality, the contributions of users in most networks are extremely unequal. For example, the number of active user addresses in Ethereum is far less than the number of unique addresses. Some users are very active, while others are not active at all.

The value of connections between users is not equal. Metcalfe's Law assumes that all connections have the same value, but in reality, different connections may have different values. Some transactions in Ethereum may involve large amounts of money, while others may involve only a small amount.

In actual networks, the growth of network size may bring marginal effects. When the network scale develops to a certain extent, each additional user may not increase the network value by the corresponding square but may decrease it, and each user's contribution to the network value will also decrease.

In summary, Metcalfe's Law has reference value for the assessment of network value, but practical applications require more data and situation analysis.

Metcalfe's Law applied to Solana

Like Ethereum, we use Metcalfe's Law to value Solana. The active user data and market cap data used are from December 16, 2020, to October 25, 2023.

Due to limited data availability with only about three years of active user and market cap data for Solana, we calculate the average value of ‘a’ over the past two years. This is obtained by dividing Solana’s daily market cap by its daily active user count and then averaging these ratios over a two-year period. The calculated value of ‘a’ is 0.76738. We apply this value of ‘a’ using the formula V=a*n^2 to estimate the market cap for all days from December 16, 2020, to October 25, 2023. We then perform a regression analysis comparing these estimated market caps with the actual market caps for each day. After removing outliers, the result R^2 value is as high as 0.8, indicating a strong correlation and the reasonableness of the chosen value of ‘a’. Therefore, we will continue to use a = 0.76738 to estimate the future value of SOL.

By October 25, 2023, the number of active addresses of Solana will be approximately 100,000 (average value of the past 30 days), and the resulting price will be $6.33 USD, according to the model. If it equals the current number of active users of Ethereum, which is 400,000, the price, according to the model, would be $101.28 USD. If it reaches 1 million active users, the price will be $633.03 USD.

Number of address(x)	Market Cap(y)	Circulating Supply	Price
100,000	7,673,800,000	1,212,240,518	6.33
150,000	17,266,050,000	1,212,240,518	14.24
200,000	30,695,200,000	1,212,240,518	25.32
250,000	47,961,250,000	1,212,240,518	39.56

Number of address(x)	Market Cap(y)	Circulating Supply	Price
300,000	69,064,200,000	1,212,240,518	56.97
350,000	94,004,050,000	1,212,240,518	77.55
400,000	122,780,800,000	1,212,240,518	101.28
1,000,000	767,380,000,000	1,212,240,518	633.03
2,000,000	3,069,520,000,000	1,212,240,518	2,532.10
3,000,000	6,906,420,000,000	1,212,240,518	5,697.24
4,000,000	12,278,080,000,000	1,212,240,518	10,128.42
5,000,000	19,184,500,000,000	1,212,240,518	15,825.65

The application of Metcalfe's Law to Ethereum valuation presents an innovative perspective, proposing that the value of Ethereum is intrinsically tied to the square of its user base. However, several aspects, such as perfect correlation assumption, quality of users, the proxy of active addresses, extrapolation of user growth, network congestion, competition from other blockchain networks, price volatility, and regulatory risks, need to be carefully evaluated as it might skew the real Ethereum network effects.

2.6 NVT Ratio

The NVT – Network Value to Transaction – ratio can be another very useful indicator for investors. It can be a useful guide to measure the network value compared to the network's usability, in this case, the transaction volume.

Created by Willie Woo, the NVT is sometimes referred to as the "Bitcoin PE ratio." Why?

Similar to the PE ratio, it correlates the value (or price) with the output of the network. NVT correlates the network value (i.e., market capitalisation) with the transaction volume (i.e., the daily volume of transactions on the network). The NVT can be low (if the price is low compared to the transaction volume) or high (if the price is high compared to the transaction volume).

This makes sense, considering that to transact, people need BTC. This higher transaction volume will also require more demand for BTC. Also, people need to pay fees to conduct transactions on the network denominated in BTC. The NVT model offers simplicity, an easy way to benchmark and compare to other cryptocurrencies and measure Bitcoin's economic activity.

2.7 NVT calculation

The Network Value to Transactions (NVT) ratio is a valuation metric for digital assets, similar to the Price-to-Earnings (PE) ratio used for equities. The NVT ratio calculates the network value (market cap) of a digital asset, which is the price of an individual coin multiplied by the total number of coins in existence. The denominator of the NVT ratio is the volume of transactions on the blockchain during a 24-hour period, reflecting its utility as a medium of exchange. NVT is expressed by the following formula:

NVT Ratio = Market value/Transaction volume

The NVT ratio should track the network value to its use closely. A high ratio may indicate either a bright future with expected growth or overvaluation. Just like high-growth companies can justify high PE ratios with expected future growth, a high NVT ratio can be justified if there's belief in the future growth of the network.

The NVT ratio can help identify bubbles in the market: when the NVT reaches around 90-95, it may indicate that the market value of the network is getting ahead of its utility. However, the NVT ratio should be used cautiously as it can often deviate from the periphery, and a smoothed average is used to get a clearer signal through the volatile fluctuations.

As of now, the NVT ratio can help to determine what happens after a price run, but its predictive power before a price run-up is unclear. Therefore, while it's an informative tool for understanding the health and potential growth of a cryptocurrency, it should be used in conjunction with other indicators and metrics for a more comprehensive view. The Network Value to Transaction is a commonly used valuation metric for different cryptocurrencies, especially L1s.

The NVT will give a ratio that helps to understand the value of the network compared to the volume transacted.

The NVT correlates the network value with the volume of the transactions on the network.

1. Collect the network value
2. Get the total transaction volume in USD over a period of time
3. Calculate the NVT = Network Value/Transaction Volume

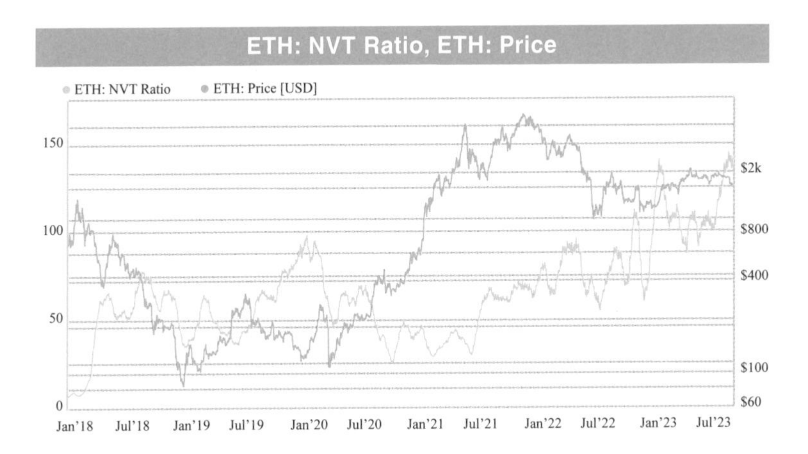

A high NVT suggests that the network is overvalued compared to the transactions, and a low NVT indicates that it might be undervalued compared to the transactions that it has.

However, in most cases, the NVT ratio doesn't account for the transaction volume that happens with tokens and that happens in L2s. If we account for this, the NVT can be much different.

If we propose the NVT adapted to tokens + L2 activity, the formula is as follows:

NVT adapted to tokens + L2 = Network Value/(Transaction Volume on L1 + L2 + token tx volume)

This approach attempts to capture a more holistic view of the economic activity in the Ethereum ecosystem, which happens not only at the L1 level but also at the L2 level, as well as other tokens that are alive on the network. This includes activity also at the level of DeFi protocols.

Estimating Ethereum's Value Using the NVT Ratio

Here, for the sake of simplicity (taking into account the presence of numerous L2 or DApps and tokens on the Ethereum network), we still use the original formula of the NVT ratio, **NVT = Network Value/Transaction Volume**, to evaluate the value of Ethereum. Therefore, the Transaction Volume used here only includes the transaction volume of Ethereum L1.

We select data from the past five years (October 2018 to October 2023) to analyse Ethereum's NVT Ratio. The lowest point of the NVT within five years was in September 2020, which is likely related to the large number of transactions brought about by the DeFi summer led by DEX activity in 2020.

According to the data on the percentage of ETH supply in smart contracts as recorded by Glassnode, we can indeed confirm that starting from September 2020, the proportion of ETH supply in smart contracts began to increase significantly and has continued until now. After entering the bull market in 2021, the NVT began to fluctuate and rise until now.

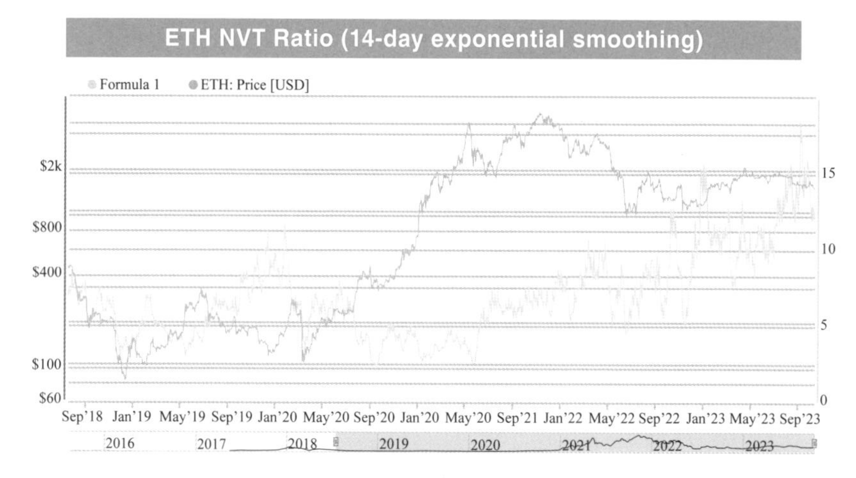

Source: Glassnode, Compiled by HashKey Capital

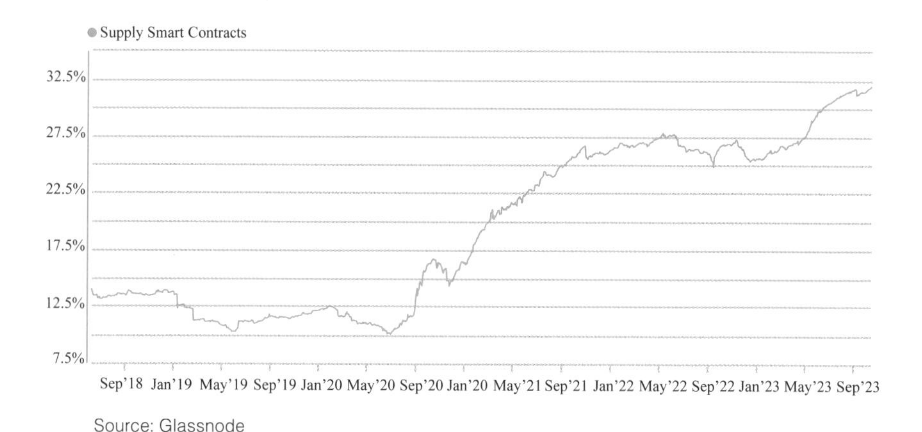

Source: Glassnode

According to Santiment data, we take the **average of Ethereum's NVT over the past five years, which is 73.64**. Then, we estimate the market value and price based on the estimated Ethereum transaction volume. The growth assumption for Ethereum transaction volume follows the growth rate calculated previously for the DCF (i.e., it assumes that the growth rates of Ethereum transaction fees and transaction volume are the same). From this, it is estimated that by 2030, the price of Ethereum will reach **51,219.96 USD.**

Year	Growth rate	Transaction Volume (Estimated)	NVT Ratio
2023 (Annualized)	-	27,870,586,558	73.64
2024	25%	34,838,233,198	73.64
2025	25%	43,547,791,497	73.64
2026	20%	52,257,349,796	73.64
2027	20%	62,708,819,756	73.64
2028	10%	68,979,701,731	73.64
2029	10%	75,877,671,904	73.64
2030	10%	83,465,439,095	73.64
Year	Market Cap	Price	
2023 (Annualized)	2,052,389,994,132	17,103.25	
2024	2,565,487,492,665	21,379.06	
2025	3,206,859,365,831	26,723.83	
2026	3,848,231,238,997	32,068.59	
2027	4,617,877,486,797	38,482.31	
2028	5,079,665,235,476	42,330.54	
2029	5,587,631,759,024	46,563.60	
2030	6,146,394,934,926	51,219.96	

Data Source: HashKey Capital Organization

Limitations

Reliability of transaction volume: The accuracy of the NVT ratio depends on accurate transaction volume data. However, due to the characteristics of the cryptocurrency market, the reliability of the transaction volume data might be challenged due to manipulation and false transactions, and the transaction volume may be affected by one-time large transactions, which may not fully reflect the usage of the asset. At the same time, this model cannot capture off-chain OTC activity and might not catch L2 activity, and it doesn't account for other factors.

Market liquidity: The NVT ratio does not consider market liquidity factors. Even if the transaction volume is large, if market liquidity is low, it may lead to instability and manipulation of transaction prices.

Market sentiment and speculative factors: The NVT ratio cannot capture the sentiment and speculative behaviour of market participants, which also have a significant impact on the price and market value of cryptocurrencies.

Solana's NVT

As for the NVT ratio of Solana, similar to the method used to calculate the constant 'a' in Metcalfe's Law mentioned earlier, we calculate the average NVT ratio for each day over the period from December 16, 2020, to October 25, 2023. This is obtained by dividing the market cap by the daily trading volume. We use this average NVT ratio for Solana to estimate its future price. The calculated NVT ratio for Solana is **29.45**. We use the same method as Ethereum's valuation and assume that the growth of transaction volume and income growth are consistent. Based on this, we make the following price prediction: according to the NVT model, by 2030, the price of SOL will reach **206.34 USD**.

Year	Growth rate	Transaction Volume (Estimated)	NVT Ratio
2023 (Annualized)	-	2,836,080,512	29.45
2024	25%	3,545,100,640	29.45
2025	25%	4,431,375,800	29.45
2026	20%	5,317,650,960	29.45
2027	20%	6,381,181,151	29.45
2028	10%	7,019,299,267	29.45
2029	10%	7,721,229,193	29.45
2030	10%	8,493,352,113	29.45
Year	Market Cap	Circulating Supply	Price
2023 (Annualized)	83,522,571,078	1,069,736,938	78.08
2024	104,403,213,848	1,098,919,989	95.01
2025	130,504,017,310	1,124,064,775	116.10
2026	156,604,820,772	1,146,812,764	136.56
2027	187,925,784,897	1,166,398,424	161.12
2028	206,718,363,413	1,188,593,832	173.92
2029	227,390,199,734	1,199,234,698	189.61
2030	250,129,219,728	1,212,240,518	206.34

Conclusion

The Network Value to Transactions (NVT) ratio is a valuable metric that provides insights into the valuation of cryptocurrencies by calculating the network value of a digital asset relative to the volume of transactions on the blockchain. This ratio can be instrumental in helping investors identify overvaluation, anticipate future growth, and potentially detect market bubbles.

However, like all metrics, the NVT ratio has its limitations and should not be used in isolation. It relies heavily on the accuracy and reliability of transaction volumes, which can be affected by market manipulations and one-time large transactions, and does not capture off-chain or Layer 2 activity. Moreover, the NVT does not account for market liquidity or the impact of market sentiment and speculative behaviour.

To address these limitations, we proposed an adapted NVT ratio that encompasses L1, L2, and token transaction volumes. This approach offers a more comprehensive view of the economic activity within a blockchain ecosystem, such as Ethereum. However, this method's effectiveness is contingent on data availability, and it may face challenges in capturing all transactional activity accurately.

Lastly, our analysis of Ethereum's NVT ratio over the past five years indicates that its market value and price can be estimated based on transaction volume growth assumptions. However, this estimation should be taken cautiously due to the abovementioned limitations.

In summary, while the NVT ratio can be a powerful tool in cryptocurrency analysis, it should be used alongside other metrics and indicators to provide a more holistic and reliable assessment of a digital asset's value.

2.8 P/S Ratio

The P/S ratio is one of the common indicators in the traditional financial field used to evaluate the relationship between a company's market value and its sales revenue. It calculates the ratio between the company's market value (i.e., price) and its sales revenue.

In the traditional finance world, the "price" in the P/S ratio refers to the market capitalisation of a company (the total value of all its stock), while "sales" refers to the company's total revenue. In the world of cryptocurrencies, we can adapt these concepts as follows:

- "Price" can be interpreted as the market capitalisation of the cryptocurrency, which is calculated by multiplying the current market price of the cryptocurrency by the total circulating supply or by the fully diluted market capitalisation.
- "Sales" is where we need to think a bit differently. A cryptocurrency doesn't generate sales or revenue in the same way a company does. Instead, we can use an equivalent measure that reflects the economic activity or utility of the network.

One possible approach is to use the total value of transaction fees on the network as a proxy for "sales." This reflects the utility and usage of the network, similar to how sales reflect the activity of a company.

A lower P/S ratio may mean that the company's stock price is undervalued, while a higher P/S ratio may mean that the company's stock price is overvalued.

P/S ratio = Company Market Value/Annual Sales Revenue

Applied to crypto, the P/S ratio can be calculated the following manner:

P/S ratio = Crypto Asset Market capitalisation / Annual Sales Revenue

We use the P/S ratio to compare the valuation of the mainstream blockchains horizontally:

	P/S ratio (fully diluted)	P/S ratio (circulating)
Ethereum (ETH)	271.97x	271.97x
Solana (SOL)	1888.89x	1401.15x
Tron (TRX)	7.63x	7.63x
BNB Chain (BNB)	4262.64x	3276.93x
Avalanche (AVAX)	1787.16x	876.85x
Arbitrum (ARB)	345.77x	44.09x
Polygon (MATIC)	1019.51x	939.93x
Optimism (OP)	347.11x	52.10x

Data source: Token Terminal, HashKey Capital organised. November 2013.

It can be seen that if we compare Solana with Ethereum, in comparison to Ethereum's decentralisation, security, ecosystem scale, and economic model, from the P/S ratio perspective, Solana's currency premium seems larger. However, Solana's value lies in better user acquisition, which allows more users to join crypto applications, especially in the DeFi and NFT fields.

In many cases, the P/S ratio might be more useful to compare the same asset across time rather than comparing different crypto assets. As we saw before, different layer 1s have different characteristics, and it's hard to normalise them in order to compare the P/S ratios between them properly. Despite that, the P/S ratio is still useful to understand the price performance of the asset in relation to its revenue and infer if it is relatively overvalued or undervalued.

Chapter 3

DAO Tokens and Utility Tokens

Many tokens represent an economic interest in the issuing business, and from an economic perspective, they might resemble equity in a business. The main difference is that in most cases, crypto tokens represent an economic interest in a decentralised business (for example, a DeFi protocol) rather than in a centralised business (for example, a bank).

In this section, we are not going to attempt to categorise tokens according to their legal status or if they are a security according to the SEC or the Howey test. Instead, we will simply look at these tokens from an economic perspective.

The nature of DAO tokens and utility tokens varies, but they often offer decision-making and voting rights, the right to use the platform with some sort of benefit, and the right to receive future distributions.

This allows us to evaluate these tokens through the lens of traditional valuation methods that include the income approach and market approach.

Method	Goal	Usability
Market approach	Evaluating the token according to its market price	Any crypto asset
Comparable tokens	Comparing qualitative or quantitative metrics of a project with competitors	Most useful for early stage projects
Quantity Theory of Money	Measuring the velocity and usefulness of a crypto asset	Any crypto asset
Market capitalisation/ TVL ratio	Correlating the market capitalisation with the Total Value Locked	Tokens related to DeFi protocols

3.1 Market approach

Tokens that are openly traded in the secondary market and that benefit from a fair amount of liquidity with updated prices against other currencies can benefit from the market approach.

In the market approach, we can simply compare the quoted token prices listed on the major crypto exchanges.

According to the efficient market hypothesis, the price of an asset reflects all the news and performance of the underlying business, i.e. it reflects the value of the asset. This would also assume that the market value equals the fundamental value. Saying this, to evaluate a token according to its price, we need to ensure that the token trades in a number of exchanges and it has enough liquidity.

Considering that the token has liquidity and trades in different places (to allow arbitrageurs to level the price), we can adopt the token price as the market value for the asset and evaluate the crypto asset from the efficient market hypothesis lens.

3.2 Comparable tokens

Comparison is especially useful in tokens that are not traded in secondary markets or crypto businesses that are still fundraising and haven't launched a token yet.

Illiquid tokens or unlisted tokens (also known as pre-TGE in the crypto community) can be compared to other tokens that trade in illiquid markets.

When comparing tokens, for example, between Project A and Project B, one needs to pay attention that, most times, project A and B are in different stages and have different levels of maturity and different levels of risk.

However, by looking at some quantitative and qualitative metrics, it is possible to infer what could be a possible valuation when comparing token A with token B. This is a similar approach to what VCs use when assessing the potential future value of a token that will be launched in the future when companies raise their seed stage rounds.

VCs often use a scorecard method to assess early-stage businesses and understand their fundamental value.

Here's a practical example:

We are evaluating a seed-stage startup with no token or equity in a secondary market. In this practical example, our investment prospect has three comparable competitors.

- Competitor A valuation: $575 million
- Competitor B valuation: $8.1 billion
- Competitor C valuation: $14.8 billion

These valuations are based on their last funding round.

The next step is to calculate the average valuation of the competitors:

Average comparable valuation = [val1+val2+valn]/n

Which, in this case, is $7.65 billion.

The average valuation of the comparable tokens will be our benchmark for this analysis.

The next step is to build our scorecard.

To build the scorecard, we first have to establish the comparison factors. These comparison factors can be, for instance, team, size of the opportunity, product, marketing, number of users/wallets, number of transactions, fees, number of developers, etc.

Once the quantitative and qualitative comparison factors are established, it's time to give weight to each of them. The weight will correspond to the importance of the comparison factor to the business.

The comparison % corresponds to how the comparison factor compares with the competitors. Let's say if the team is as talented as the competitors, we would attribute it a 100% score. If the product is roughly 25% better than the competitors, we will attribute it a 125% comparison factor and so on.

Comparison factor	Weight %	Comparison %	Factor (WxC)
Team	30%	100%	0.3
Size of the opportunity	25%	150%	0.375
Product	15%	125%	0.187
Number of users	10%	120%	0.12
Marketing/sales/ partnerships	10%	100%	0.1
Need for additional investment	5%	100%	0.05
Other factors	5%	100%	0.05
SUM	-	-	1.182

We then multiply the weight by comparison to get the factor. After summing up the factors, we will get the final factor, which, in this case, is 1.182.

We can now multiply the factor by the average valuation: 1.182 * $7.65 billion = $9.04B.

This value could represent a potential valuation for the company in the future and help investors rationalise the potential multiples on the investment being analysed.

The table above is merely an example of the comparison factors that could be used. Depending on the project/token, investors can include other variables that allow them to compare the target company with their competitors. We have listed below other factors/variables that could eventually be used. The only rule is that the comparison factors that the investor uses are available and clear for both the target project and competitors in order to make fair comparisons.

Other variables to be included:

- Business model
- Management
- TAM, SAM, SOM
- MAU
- Sustainable competitive advantage
- Attractive industry
- Risks and threats
- Balance sheet
- Active wallets/transactions
- Developer ecosystem, etc.

Conclusion

The comparison of tokens, particularly those that are illiquid or unlisted, is crucial in the crypto investment landscape. This comparative analysis allows investors to infer the potential valuations of these tokens by observing both quantitative and qualitative metrics, akin to the approach utilised by venture capitalists (VCs) in assessing the potential future value of a token.

This method provides a comprehensive and systematic approach to valuing tokens in illiquid or pre-launch stages, thus aiding in more informed investment decisions. It can, however, also be used in liquid tokens to compare different projects and help investors during their decision process.

The scorecard method is flexible and can accommodate a wide range of comparison factors. Depending on the project or token, investors can include various variables for comparison. This approach is transparent, provided the comparison factors are clearly defined and available for both the target project and competitors.

Apart from the factors mentioned in the example, other variables like business model, management, market size (TAM, SAM, SOM), monthly active users (MAU), competitive advantage, industry attractiveness, risks, balance sheet, active wallets/transactions, and developer ecosystem can be included in the scorecard, making the process more robust and comprehensive.

Forecasts are hardly accurate

Generally speaking, it's hard to forecast returns on investment across most asset classes. Bonds, stocks, commodities, art, and even the property market are very hard to forecast. Most times, investors use historical returns to predict future returns, but this method assumes that patterns and trends of the past will continue into the future, and again, this method often fails.

In other words, any investment whose expected return is above the risk-free rate has what we call a risk premium and consequently has a degree of hard-to-predict volatility.

Investing in new technologies, new sectors, and young companies naturally carries a substantially higher risk premium. It's thus the role of VCs to arbitrage this risk premium and assess the risk/reward metrics for these investments. Individual investors investing in early-stage projects (crypto or not crypto) should be aware that there's a direct correlation between high premiums and volatility.

According to findings from the European Investment Fund (EIF) about the performance of early-stage venture capital (VC) investments, approximately 57% of these investments yield a multiple on money (MoM) of less than 0.25 times.

According to the data, 80% of the ICOs have failed and are no longer operational, and 30% of the VC-backed startups fail, while close to 70% of the startups never reward the investors with a profitable exit.

Saying this, we hope that the valuation metrics presented in this chapter (and in the book) help investors navigate the space with more clarity and increase the chances of getting above-average returns.

Utility tokens in distributed peer-to-peer organisations

The intrinsic value of utility tokens will always be tied to the underlying platforms and the goods/services that the platform offers.

However, that same utility token is used as an economic incentive and ownership tool for the platform operator and investors.

This means that, unlike stocks in a company, utility tokens have unity. Unity, in this case, means that the same token is used as utility but also economic extraction value and governance (the two latter resemble a stock in a company).

Saying this, utility tokens tend to be valued according to the fiat value that they provide in the underlying platform. Consequently, we can compare the utility token price used to buy the services in the platform with their Web2 alternatives.

For example, users use Filecoin tokens to pay for decentralised storage space. There are many other cloud providers like AWS that provide a similar service. Saying this, we can measure and compare the GB/Month cost of the services.

3.3 QTM – Quantity Theory of Money

As we saw before, utility tokens can be seen as a medium of exchange for certain users and can be used as "legal tender" within their respective protocols. They create small economies that can be compared to larger fiat economies, enabling us to look at utility tokens through a QTM lens.

The Quantity Theory of Money (QTM) is an essential concept in economics that describes the relationship between the supply of money in an economy and the level of prices of goods and services.

The QTM formula can be expressed the following way:

$$M*V = P*Y$$

M: Monetary supply and corresponds to the entire token circulation supply.

V: Velocity of money, or the average frequency with which a unit of money is spent.

P: Price level

Y: Volume of all the goods/services transacted. Y is sometimes referred to as T.

Applying QTM to utility tokens

We can apply the QTM to utility tokens to calculate their monetary supply, the velocity of money, price level or volume transacted.

Money supply

When we are dealing with a traditional economy, the money supply is determined by central banks. This includes different measures that are expressed as M0, M1, M2 and M3 and represent the money in circulation, deposits, cheques, securities, funds, etc.

Fortunately, in the token economy, it's easier to calculate the money supply.

In most cases, utility tokens have a fixed money supply. Typically, this supply is set by the developers at the start of the project. In some cases (although rarer), the tokens might not have a hard cap.

As part of the tokenomics of the project, it is important to determine the circulating supply and the total supply. While in some cases the circulating supply and the total supply are exactly the same, some projects will have a circulating supply that represents the money supply in the short term and the total supply that represents the long term total money supply (typically, the circulating supply and total supply should converge).

Money velocity

Money velocity can be easily observed by looking at reported exchange activity and on-chain data.

We can benefit from the fact that there's widely available data (on-chain and off-chain) regarding transactions and velocity, which is a huge improvement when compared to calculating the money velocity of traditional fiat currencies.

Volume of goods

The volume of goods would, in a fiat economy, correspond to the GDP of the country. In a token economy, the volume of goods corresponds to the total value of services provided by the protocol in a given period.

Calculating the volume of goods for a cryptocurrency or token is challenging. Our suggestion is to:

How do you calculate the market size? One methodology could be the fees collected over a period of time or in perpetuity. Another methodology could be simply assigning the total market capitalisation of the token as the volume of goods.

Typically, a high velocity for a token suggests that it has a strong demand and is more likely to have success in the long term.

Price level

Knowing the variables mentioned before, we can calculate the price level for the token. This price can then be compared to historical secondary market prices for the token, denominated in fiat terms.

Note that the price level corresponds to the average price of goods and services in the token economy and not necessarily the token price.

Practically speaking, the price level corresponds to how much a token unit can purchase. If the price level rises, fewer goods are bought. This can also be referred to as inflation.

$$P = MV/Y$$

M: Monetary supply and corresponds to the entire token circulation supply.

V: Velocity of money, or the average frequency with which a unit of money is spent.

P: Price level

Y: Volume of all the goods/services transacted. Y is sometimes referred to as T.

The formula above can be used for any token or cryptocurrency. In the example below, we use Bitcoin data for the sake of the example:

Plugging these values into the equation P = (M * V)/Y:

P = (18.5 million BTC * 2)/25 million transactions
P = 1.48 BTC per transaction

This suggests that on average, each transaction involves around 1.48 Bitcoins. In fact, if we convert this amount of BTC to its dollar equivalent, we will approximately have the average transaction size on the Bitcoin network.

Remember, in reality, these numbers can fluctuate greatly due to factors like changes in the demand for Bitcoin, speculation, regulatory changes, and many other factors. This example is a simplification and should be taken as illustrative rather than representative of real-world dynamics.

Applications of QTM for utility tokens

To apply QTM, we can eventually add a discount rate similar to the one used in the CAPM section.

QTM can show us how small economies – in this case, token economies – react to the different variables in the QTM model. It also offers a benchmark to compare token prices with market prices and be a proxy for token valuation.

Thoughts: According to Milton Friedman, M and V should not be constants as there are many variables to take into account. "Velocity of money not as a constant, but rather as a variable determined by a small number of independent factors."

It is also a difficult task to apply discount rates according to CAPM to startup projects, as it is highly subjective to provide discount rates to any startup.

- Startups don't have historical data
- Highly unpredictable
- Variable cash flows and sometimes no cash flows
- Lack of liquidity

One could try to predict future transaction token price as P = [M * V]/Q using predictions on token supply, velocity and ecosystem growth. But again, this model might be leaving many variables out of the equation.

Note that although velocity can be linked to the value of a utility token, in some cases, reduced velocity through certain mechanisms can also have a positive impact on the price. Projects can artificially decrease velocity by introducing profit-share mechanisms, staking incentives, burn mechanisms, and gamification tactics to incentivise long-term token holding.

3.4 Market capitalisation/TVL ratio

The Market capitalisation to TVL ratio allows us to compare different projects within the same sector/use cases. The ratio can be used to compare across different layer 1 projects or different layer 2 or different DeFi DEX, etc.

The TVL represents the Total Valued Lock, that is, the funds being locked or, deposited or staked in a protocol. An increase in TVL indicates that more capital is being deposited, which means higher usage of the project. In fact, the TVL can be seen as a similar metric to a bank AUM.The higher it is, the higher the capacity of that bank (in this case, the DeFi protocol) to lend assets.

The Market Cap to TVL can be extremely useful as it gives investors a rational metric that correlates the price (market cap) with the actual usability of the protocol measured by the TVL. The lowest the ratio is, the more underpriced the asset is in comparison to the TVL.

The TVL is calculated by taking the total amount of tokens locked in the protocol and multiplying it by the price of the token.

Market capitalisation to TVL: Market capitalisation/
Total value locked

Name	Token price	Market cap
Aave	$96.30	$1,410,000,000.00
Compound	$51.50	$414,000,000.00
Venus	$6.70	$106,000,000.00
Name	TVL	Market cap to TVL
Aave	$5,774,000,000.00	0.24
Compound	$2,237,000,000.00	0.19
Venus	$680,000,000.00	0.16

Market cap to TVL interpretation: the ratio measured the relative size of the market cap with the TVL. In simple terms, similar to a P/E ratio, a higher Market Cap to TVL means that the price is more expensive when compared to the TVL. This might translate into different interpretations for the investors. A high ratio can mean that the asset is overpriced or that the investors have high expectations for the future growth of the protocol. On the other hand, a low ratio might be interpreted as an underpriced asset or an asset that is less desired by investors when compared to its peers.

Name	Price Jan. 23	Mcap/TVL Jan. 23	Price Dec. 23	Mcap/TVL Jan Dec. 23	Performance
Radiant	$0.04	0.04	$0.20	0.27	400%
Maker	$518.00	0.07	$1,480.00	0.17	186%
Venus	$4.10	0.08	$6.80	0.15	66%
Curve	$0.50	0.1	$0.60	0.26	20%
Marinade	$0.06	0.12	$0.20	0.11	233%
Liquity	$0.60	0.13	$1.50	0.20	150%
Compound	$31.90	0.14	$53.30	0.16	67%
Lido	$1.10	0.15	$2.50	0.11	127%
Aave	$52.00	0.2	$97.00	0.25	87%
Frax	$4.60	0.25	$7.00	0.63	52%
dYdX	$1.20	0.41	$3.20	1.70	167%
RocketPool	$21.00	0.69	$28.70	0.23	37%
GMX	$41.50	0.75	$48.60	0.81	17%
Uniswap	$5.40	1.23	$6.20	1.25	15%

The table above shows 14 different DeFi protocols, their price, and market cap to TVL ratio in Jan. 2023, as well as the price in Dec. 2023, the TVL and 11-month performance. We can observe that the best-performing tokens had a low market cap to TVL. Radiant, for example, had a market cap to TVL of 0.04 in Jan. 2023 and returned 400% over the following 11 months. On the other end, Uniswap had the highest market cap to TVL, scoring 1.23, and had the worst returns in the group: a measly 15%.

Across these 14 protocols, the seven lowest market cap to TVL (with a ratio under 0.15) returned 160% on average, while the seven highest market cap to TVL (above 0.15) returned 73% on average.

Combined with other metrics, the market cap to TVL ratio can help guide investors by allowing them to compare the token price (or market capitalisation) with the TVL across time and across projects in the same sector.

Conclusion

Despite the Market Capitalisation to Total Value Locked (TVL) ratio's potential usefulness, it's important to note some inherent limitations.

The TVL is not always indicative of the protocol's utility or success. It can be inflated by liquidity mining programs, where token rewards incentivise users to deposit assets, which can artificially inflate TVL. Once these incentives are removed, TVL might decrease rapidly.

Additionally, the TVL does not account for user diversity. A high TVL could result from a few large players, which would imply a concentration of assets and potential manipulation rather than widespread adoption.

It is important to say that the Market Cap to TVL ratio doesn't account for differences in risk between protocols. Some protocols may have a lower ratio due to higher perceived risk rather than being undervalued.

Finally, the ratio might not reflect the future potential of a protocol. A protocol with a high ratio might be pricing in future growth and potential, not just current TVL.

Despite these criticisms, the Market Cap to TVL ratio provides a meaningful metric for comparing the price of a project to its current level of usage, as measured by TVL.

The example provided, comparing 14 different DeFi protocols, underscores how this ratio can potentially identify undervalued projects.

The Market Cap to TVL ratio, while not perfect, offers a valuable tool for investors seeking to make informed decisions in the rapidly evolving DeFi landscape. It encapsulates the intersection of market perception and protocol usage, providing a snapshot of how the market values a protocol relative to the assets it has attracted.

Chapter 4

A Framework for DAO Token Valuation

DAOs are organisations composed of token holders that have a common purpose. There are some overlaps between DAO tokens and utility tokens, considering that a DAO token often offers governance powers and utility in the underlying protocol/service. This is very common, for example, in DeFi protocols, where the same token represents governance over a DAO, a currency for payment in the platform, or some kind of benefit like boosting a reward or fee discount.

Saying this, in some cases, we can also evaluate DAO tokens with the same token valuation methods used for utility tokens.

Some projects adopt a corporate-DAO or foundation-DAO structure where the company or foundation acts as the software house for the underlying DAO protocol. The DAO allows the users and community to propose and vote on proposals via token voting.

Typically, a DAO is democratic, and any token holder can propose a new feature or change on their governance forum. Ideas are discussed with the community and put to voting. The token holders will vote, and the weight of their votes will correspond to the proportion of the tokens they hold. If proposals form a quorum and are passed, the company/foundation moves forward with developing and implementing the proposal.

It's also important to say that in most cases, the product/service provided by the DAO is fully automated via smart contracts, which are decentralised on the blockchain. On the other hand, the responsibility of the company/foundation is to be a service provider that maintains/upgrades/fixes the middle layer, smart contract and front end of the protocol. The company/foundation might also own the IP of the brand/software. In this section, we will focus on the valuation of the DAO and will exclude the company/foundation associated with it, as it can be seen as a mere vendor.

DAO valuation can be done using two different approaches: fundamental valuation and comparable analysis.

Note that we are about to provide a generic framework for DAOs, as DAOs can have many different purposes and categories: from protocol DAOs, grants DAOs, social DAOs, venture DAOs, collectors DAOs, etc.

When using the fundamental approach and the comparable analysis, it's important to compare the DAO being evaluated with other DAOs in the same category.

Fundamental approach: value the DAO token according to the fundamentals of the token, utility, tokenomics and token value accrual.

Comparable approach: value the DAO by using metrics to compare it with other DAOs in the same sector.

In this section, we will be covering fewer valuation methodologies. This is not because of the lack of methodologies for DAO token valuation but because we have covered most methodologies already in this book, and there's no need to repeat it.

Method	Goal	Usability
Net Current Asset Value Per Token - NCAVPT	Calculating the liquidation value of a crypto asset or DAO	DAOs and projects with assets and a treasury

4.1 Voting premium

Does voting have value?

Similar to common stocks and preferred stocks in traditional finance, voting does have a premium. Voting has power over the future of the project, and consequently, it has an intangible value that, although very hard to measure, it still exists.

In the crypto space, there's been a fair amount of discussion on whether voting has any impact on the token value. As per Buterin's (2022) argument, it is important to separate governance rights from token value in order to establish a just and transparent token economy that benefits all stakeholders.

4.2 Net Current Asset Value Per Token (NCAVPT)

Benjamin Graham popularised NCAVPS - Net Current Asset Value Per Share - to gauge the attractiveness of a stock. This is a metric that is calculated by taking the current assets, subtracting total liabilities, and dividing it by the total number of shares.

The Net Current Asset Value resembles the liquidation value, which is mostly used when a company is facing bankruptcy and or is being wound down. The estimation of a company's minimum value is reflected in the liquidation value. This value is calculated by considering the net realisable cash that would be obtained from selling all assets and paying off liabilities. For stockholders, this value represents the leftover money available for distribution once all debts and obligations have been fulfilled. The liquidation value ignores any potential future revenue or earnings of the company.

In this section, we have adapted the Net Current Asset Value Per Token (NCAVPT) to allow investors to compare the NCAVPT value of a DAO with its token price. Most DAOs might be valued according to their NCAVPT.

DAOs such as investment DAOs and collector DAOs, unlike DeFi or protocol DAOs, typically don't generate cash flows. Instead, the goal of this kind of DAO is to be a collective vehicle for the token holders to invest in other projects (just like an investment syndicate or VC) or to collect art (both physical art and NFTs). *Other DAOs, like the Noums DAO, also fall into this category.*

The NCAVPT of a DAO can give investors important data. For company shareholders, the NCAPS liquidation value symbolises the residual funds available for distribution after settling all debts and obligations. This valuation method inherently disregards any prospective future revenue or earnings of the company. We can replace "company" with "DAO" and "share" with "token", and we will get the same.

In this type of DAO, the value of the DAO might be close to the liquidation value.

NCAVPT = (Current Assets - Total Liabilities) ÷ Token Supply

In traditional securities, the liquidation value equals the total assets subtracted from the liabilities. However, in most cases, DAOs have very few or no liabilities.

Valuing the assets of the DAO – Assuming that the assets are liquid, the assets of a DAO are digital assets such as fungible tokens and NFTs. These are usually easy to value at market prices.

Subtracting liabilities of the DAO – Although not common, DAOs might have liabilities such as loans on DeFI protocols.

Subtracting the assets from the liabilities will be the liquidation value of a collector DAO or investment DAO.

Token	Assets	Circulating supply	NCAVPT	Current Price
Optimism	$4,500,000,000	911,295,000	$4.94	$2.10
Mantle	$2,500,000,000	3,132,673,000	$0.80	$0.59
Arbitrum	$4,000,000,000	1,275,000,000	$3.14	$1.08
Uniswap	$2,800,000,000	588,187,016	$4.76	$6.27
Gnosis	$1,800,000,000	2,589,588	$695.09	$233.00
dYdX	$685,500,000	183,765,523	$3.73	$2.80
ENS	$669,100,000	30,326,169	$22.06	$9.50

Token	Assets	Circulating supply	NCAVPT	Current Price
The Graph	$541,100,000	9,322,894,087	$0.06	$0.16
ReseachCoin	$464,500,000	76,216,481	$6.09	$0.50
Lido	$368,100,000	889,560,019	$0.41	$2.27
Stargate	$237,800,000	204,338,417	$1.16	$0.53
Frax	$224,700,000	75,472,333	$2.98	$8.96
Decentraland	$196,800,000	1,893,095,371	$0.10	$0.49
Orbs	$189.600.000	3,167,720,359	$0.06	$0.04
YGG	$187,000,000	284,903,702	$0.66	$0.36
Merit Circle	$150,000,000	322,101,826	$0.47	$1.84

Source: HashKey Capital. December 2023

As we can see, according to the NCAVPT in this table, some tokens have a Net Current Asset Value higher than the current price, while others are under the current token price. In other words, some tokens trade at a lower value than the Net Current Asset Value Per Token (i.e. they are underpriced according to this metric) while others trade at a higher value (i.e. they are overpriced).

Conclusion

The liquidation value and NCAVPT provide investors with a gauge of the DAO's token minimum worth. It can also be useful in a case where the DAO token price is trading at such a low price that the market capitalisation is lower than the liquidation value or the NCAVPT (net asset value) is higher than the token value. Moreover, it can also serve as a benchmark for comparing investment opportunities across various DAOs. However, like all valuation methods, it should be utilised in conjunction with other metrics and indicators to formulate a more holistic view of the DAO's valuation.

Investors need to take into consideration that the NCAVPT doesn't look at the business model of the DAO. Some DAOs might not need a treasury at all to function. In this case, the Net Current Asset Value would be low, but that's simply because a treasury is not needed in that specific use case.

Another limitation of the Net Current Asset Value is the fact that, in some cases, the DAO treasury is partially composed of its own token. Consequently, changes in the token price will have a recursive effect on the DAO treasury and affect the use of the NCAVPT.

4.3 Transparency and governance premium

We also want to use this section of the book to talk about the transparency premium and the governance premium.

First, blockchains are natively transparent and permissionless, meaning that it is way easier to verify on-chain data, such as the number of users, revenues, treasury, TVL, etc. Additionally, most blockchain-based projects are open source, which further increases the transparency.

Why is this important?

Typically, in the traditional equity world, investors only have access to data through the company's annual or quarterly reports. These reports have months of delay between the facts and the moment when they are released. Additionally, they rely on accountants and external auditors who can cook the books. Although relatively rare, accounting fraud has generated huge losses for investors in the past.

On the other hand, tokens and cryptocurrencies provide users with real-time, transparent, immutable, and tamper-proof data. This allows investors to collect data regarding the project at any point in time without the need to wait for a quarterly report and without the need to trust an auditor.

This fact should give cryptocurrencies and tokens a premium.

Another important point is the governance premium. Traditionally, companies can have two types of stocks: preferred stocks and common stocks. Typically, only the second, common stocks, give investors voting rights. This generates two categories of investors: the ones who can't vote and others who can vote. It also generates a premium, where, many times, common stocks trade at a higher value than the preferred stock. This shows that voting right has a premium for investors.

In the same way, the fact that many blockchain projects, protocols, and DAOs have a token that token holders can use to vote should translate into a gaugeable premium for the token.

Chapter 5

STO Valuation

Security Token Offering (STO) can be viewed as a hybrid approach, combining the innovative aspects of Initial Coin Offerings (ICOs) with the regulatory compliance and asset backing found in traditional Initial Public Offerings (IPOs). STOs are characterised by the issuance of security tokens, which represent ownership in real-world assets/securities such as equities, bonds, or other financial instruments. Consequently, the valuation of security tokens is rooted in established financial valuation methodologies.

Traditional securities always back an STO, and the STO product can be seen as the technology infrastructure used for the distribution and record of the asset: instead of merely using traditional institutions for asset data and settlement, it uses the blockchain.

Saying this, an STO valuation should be the same as any other security that is traded using more traditional railways, such as a regular stock exchange.

Although an STO can be seen and evaluated as a traditional security, in this section, we will primarily use three methods for STO valuation: Income-based Methods, Relative Valuation Method, and Cost approach.

Investors and analysts can access three comprehensive methods for estimating the value of security tokens issued during an STO. The selection of a particular valuation method depends on factors such as the nature of the underlying assets, the industry, and the specific investment objectives. The accuracy and reliability of STO valuations are crucially dependent on regulatory compliance and transparency, as these factors directly impact investor trust and confidence in this evolving financial landscape.

Method	Goal	Usability
Income-based Approach	Similar to DCF, it aims to evaluate a security based on its future income.	STOs with equity/security as an underlying product.
The Relative Valuation Method	Compare similar asset in the same industry	Most STOs
Coat Approach	Estimate the value based on the principle of substitution	STOs with real-world assets (such as property) as an underlying asset.

Considering that an STO token represents an underlying security, and the valuation of such a product would require a book in and of itself, we will have a more theoretical and short approach in this section of the book.

5.1 Income-based Approach

The Discounted Cash Flow (DCF) is the most widely used income-based method, and it is a widely used valuation approach that is suitable for estimating the value of security tokens in Security Token Offerings (STOs). DCF is based on the principle that an asset's worth is determined by its ability to generate future cash flows for investors. This method can be used in various scenarios, including valuing companies, pricing bonds, and assessing the worth of real estate investments. Therefore, DCF is a reliable tool for valuing STO underlying assets.

We discussed the DCF approach to crypto assets earlier in the book, but we will also break it down for securities (STO).

For an STO, four key components instead of three due to traditional DCF require consideration: cash flows (CF_i), the discount rate (r_i), the number of periods (n) and STO discount. Here's a more detailed exploration of these elements:

1. Cash Flows:

The term "cash flows" encompasses various sources of earnings or dividends associated with the asset. These can include dividend payments, interest revenue, proceeds from asset sales, free cash flow, or residual income. The specific cash flow components depend on the type of assets and the interests of the investors. To employ this method effectively, an analyst must conduct comprehensive research on the business model and make well-founded assumptions regarding growth rates to estimate future cash flows. In practice, to simplify the assumptions for forecasting, cash flows are often divided into two categories: forecasted cash flows and terminal cash flows.

2. Discount Rate:

The discount rate represents the rate of return expected by an investor for allocating their capital to the project. It signifies the minimum return an investor would accept in lieu of investing in alternative opportunities. It's important to note that a higher discount rate results in a lower present value for future cash flows, and conversely, a lower discount rate increases the present value. The selection of the appropriate discount rate is pivotal, as it should reflect the risk profile, uncertainty, and time value of money associated with the project. Various methods can be employed to estimate the discount rate, including the cost of capital and market comparables, among others.

Estimating the discount rate is a crucial step in determining the feasibility of a project or company. One commonly used method to estimate the discount rate is the cost of capital approach. This approach calculates the weighted average cost of capital (WACC) by taking into account the rates of return that need to be paid to the investors, including both debt and equity holders.

To calculate the WACC, the cost of each source of capital is multiplied by its proportion in the capital structure and then added up. The cost of debt can be estimated by using the interest rate on the debt, adjusted for taxes. On the other hand, the capital asset pricing model (CAPM) can be employed to estimate the cost of equity. The CAPM considers various factors, such as the risk-free rate, the market risk premium, and the beta of the project or the company, to estimate the cost of equity.

w_d: the weight of debt
w_e: the weight of equity
r_d: required return of debt
r_e: required return of equity
t: tax rate

To determine the capital structure weights, we need to calculate the percentage contribution of each source of capital. For instance, if our company's market capitalisation or equity value is $200 million, and the net debt balance is $80 million, then by adding these two values, we can calculate the company's total capitalisation, which is $280 million.

Using this information, we can determine the relative weights of debt and equity in the company's capital structure. The equity weight is calculated as 71%, and the debt weight is calculated as 29%.

The cost of debt can be found at the IRR of the corporate bond of the company or a comparable company, and considering the effect of tax shield, we must calculate the after-tax cost of debt:

- Estimating the cost of equity is more complex. We can use CAPM to calculate r_e.
- Rf refers to the risk-free rate of return, which can be approximated by a US government bond with a duration that matches the investor's expected timeframe for owning the stock. The 5-year T-bill is a good proxy for this. At present, the 5-year T-bill yields 1.7%, and the 10-year yields 2.2%. Thus, a 2% risk-free rate is a reasonable approximation.
- B (Beta) represents the sensitivity of the expected stock return to the market return, and historical data is used to estimate it. Mathematically, it is calculated as the covariance of the historical return of a particular stock and the market divided by the variance of the market. To determine B, one should examine the beta of similar public stocks. For public SaaS companies, the beta is currently around 1.3.

- Rm denotes the market rate of return or what investors anticipate the market will return. Over the last decade, the public markets have returned approximately 8% per year, which is a reasonable expectation. However, there may be differing opinions, such as if the 5-year rate of return is much higher. In the case of private companies, one would expect a much higher rate of return.

3. Forecast and Terminal Periods:

The forecast period is the span during which a company can reasonably predict its future performance and plan its operations with a high degree of confidence. In a traditional business context, this forecast period typically ranges from 8 to 15 years. In contrast, the terminal period extends beyond the forecast horizon and can theoretically stretch into infinity. When evaluating the terminal cash flow, valuation professionals calculate the discounted sum of cash flows for all years beyond the initial forecast period. The terminal growth rate, denoting the annual increase in cash flows, is a critical factor in this context, as it influences the valuation outcome.

4. STO discount:

So far, points 1. 2. and 3. of the DCF calculation would be done using the traditional DCF model. However, as a final step, we add the STO discount.

$$DCFValueSTO = DCFValue * (1-d)$$

In essence, the DCF method can be a robust and widely employed approach for valuing security tokens in STOs, providing a structured framework for assessing the intrinsic value of these assets based on their potential to generate future cash flows. Accurate valuations require meticulous research, well-informed assumptions, and thoughtful consideration of cash flows, discount rates, and forecast versus terminal periods.

Finally, here's the complete DCF Valuation Formula with STO Discount:

$$DCFValueSTO = \left(\sum_{i=1}^{n} \frac{CF_i}{(1 + WACC)^i} + \frac{CF_{n+1}}{(WACC - g)} \times \frac{1}{(1 + WACC)^n} \right) \times (1 - d)$$

This formula represents the complete integration of the DCF method for valuing the underlying assets in an STO. It incorporates the present value of the cash flows throughout the forecasted period, the terminal value, the calculated discount rate through WACC, and any additional STO-specific discount factors.

5.2 The Relative Valuation Method

The Relative Valuation Approach, also known as "comps" or "comparable company analysis," is an effective method that assesses a company's value by comparing it to similar firms or precedent transactions. Comparable companies are those that share a similar business model, making them suitable benchmarks for evaluating the financial worth of the target company. When selecting precedent transactions, it's crucial to choose those that closely resemble the company's specific circumstances.

The fundamental premise behind relative valuation is that similar assets should have similar market values. This approach avoids the need to predict future cash flows while still adhering to the assumption of a going concern. Relative valuation offers a unique advantage in that it is firmly grounded in "reality" as it derives value from actual trading prices readily observable in the market.

A valuation multiple, a key component of relative valuation, comprises two parts:

- **Numerator**: Typically, this represents a value measure, such as Enterprise Value (EV) or Equity Value. These value measures encapsulate the total value of the company's operations.
- **Denominator**: Denominator usually involves a Value Driver, which can be a financial metric (e.g., EPS, EBITDA) or an operating metric (e.g., number of users). The choice of the denominator depends on the specific characteristics of the industry and the assets being evaluated.

There are several common types of relative valuation ratios, including

- **Price-to-Earnings (P/E) Ratio**
- **Price-to-Book Value (P/B) Ratio**
- **EV/EBITDA Ratio**
- **Price-to-Cash Flow Ratio**
- **Price-to-Sales Ratio**

The selection of the appropriate valuation multiple depends on the nature of the underlying assets and the industry context.

It's important to recognise that no two assets are exactly identical, and differences must be taken into account in any relative valuation analysis. This method is often used as a validation tool to corroborate the results obtained through the Discounted Cash Flow (DCF) approach, providing an additional layer of confidence in the valuation outcome. When applied to a large sample of analogous companies and combined with various valuation methods, the correct estimate should typically fall within the range of valuations generated by each methodology. This convergence of valuation results helps enhance the overall reliability of the assessment.

5. Cost Approach

The Cost Approach is grounded in the fundamental economic principle known as the "principle of substitution." This principle posits that rational investors will not pay more for a property than they would for a substitute property of equal utility. the cost approach is especially useful for STOs related to real-world assets.

The Cost Approach involves two key components: **reproduction cost** and **replacement cost**. Each of these facets plays a crucial role in determining the value of a property or asset.

1. Reproduction Cost:

Reproduction cost refers to the total expense associated with precisely duplicating an asset or property using the same materials and specifications as the insured property, considering current market prices.

2. Replacement Cost:

Replacement cost, on the other hand, represents the amount of money a business or investor must expend to replace an essential asset with a new one with the same or higher value. This is a practical assessment, considering the current market conditions and the availability of equivalent assets. Replacement cost is often used in cases where an asset might need to be replaced due to wear and tear or obsolescence.

The Cost Approach is particularly well-suited for various scenarios, including

- **Special Use Properties:** Properties that have unique characteristics or purposes, making it challenging to find direct comparables through other valuation methods.
- **New Construction:** When valuing newly built structures where there may be limited market data available for comparison.
- **Insurance:** Calculating the coverage needed to adequately insure a property or asset, ensuring that the policy reflects the cost of replacement in case of damage or loss.

- **Commercial Property:** Assessing the value of commercial real estate assets, especially when the property has specific features that distinguish it from other properties in the market.

The Cost Approach provides valuable insights into the intrinsic value of assets based on the costs involved in reproducing or replacing them. It is particularly useful when market data is scarce or when valuing properties with unique characteristics. However, it's essential to consider that this approach does not account for factors like market demand, location, or economic conditions, which may influence the market value differently. Therefore, it is often used in conjunction with other valuation methods to provide a comprehensive view of an asset's worth.

Conclusion

Security Token Offerings (STOs) present a sophisticated blend of the innovation seen in ICOs with the structured compliance of IPOs. By issuing security tokens that are tied to actual financial instruments, STOs offer a tangible connection to the real economy, with valuations grounded in traditional financial analysis methods. The three valuation methods detailed—Income-based, Relative Valuation, and Cost Approach—provide a versatile toolkit for investors and analysts to determine the fair value of these digital assets.

The Income-based Approach, with its cornerstone method of Discounted Cash Flow (DCF), provides a detailed framework that factors in the future cash-generating potential of the asset, appropriately adjusted for risk and time value of money. The Relative Valuation Method offers a comparative perspective, leveraging the market values of similar entities to gauge worth. The Cost Approach is particularly effective for assets where comparable market data is scarce, focusing on the costs of reproducing or replacing the asset.

Each method has its distinct application scenarios and is selected based on the asset's nature, industry context, and the specific objectives of the investment. This theoretical overview is very concise but underscores the importance of methodical research, regulatory adherence, and market transparency in achieving reliable STO valuations. With STOs, we witness a promising convergence of traditional finance and blockchain technology, heralding a new era where asset and digital finance may significantly transform investment practices.

Chapter 6

NFT Valuation

NFT stands for "Non-Fungible Token." It is a type of digital asset that represents ownership or proof of authenticity of a unique item or piece of content on a blockchain network. Unlike cryptocurrencies like Bitcoin or Ethereum, which are fungible and can be exchanged on a one-to-one basis, NFTs are unique and cannot be exchanged on a like-for-like basis since they are assigned unique identification codes and metadata that distinguish them from other tokens.

To start, let's have a brief overview of the overall NFT market. As of December 31, 2023, the market capitalisation of NFTs stands at $7.7 billion, approximately 2.6% of the market capitalisation of Ethereum.The NFT market is still very young and relatively small.

Source: NFTGO

Despite a decrease of 46.75% in the number of traders compared to a year ago, buyers decreased by 56.1%, and sellers decreased by 39.04%, the number of holders has continued to grow, with a 33.6% increase compared to a year ago.

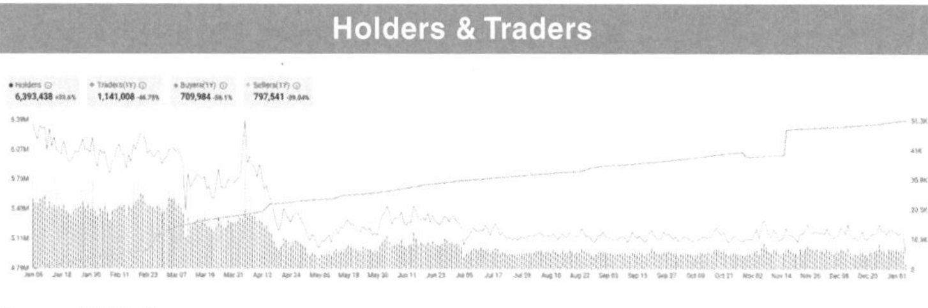

Source: NFTGO

While in 2020, most NFT use cases are associated with digital art, just a couple of years after, there are use cases where NFTs represent other assets, such as PFPs, domains, loans, identities, contracts, DeFi liquidity positions, tickets, music, etc.

In terms of asset types, as of December 31, 2023, the PFP (Profile Picture) category continues to dominate the market. PFPs lead in terms of sales volume, market capitalisation, and trading volume, surpassing other categories. The market capitalisation of PFP NFTs accounts for approximately 60% or more of the overall NFT market, making it a significant component of the current NFT market.

Source: NFTGO

According to NonFungible, the sales of NFTs were very active in the second half of 2021 and the first half of 2022. However, as the web3 market cooled down in the second half of 2022, the sales of NFTs experienced a significant decline. The NFT market has been entering a phase of "de-bubbling" since then. Also, OpenSea's position as the leading NFT trading platform has been increasingly threatened since 2022. As of early 2023, Blur has emerged as the new market leader with a zero transaction fee policy.

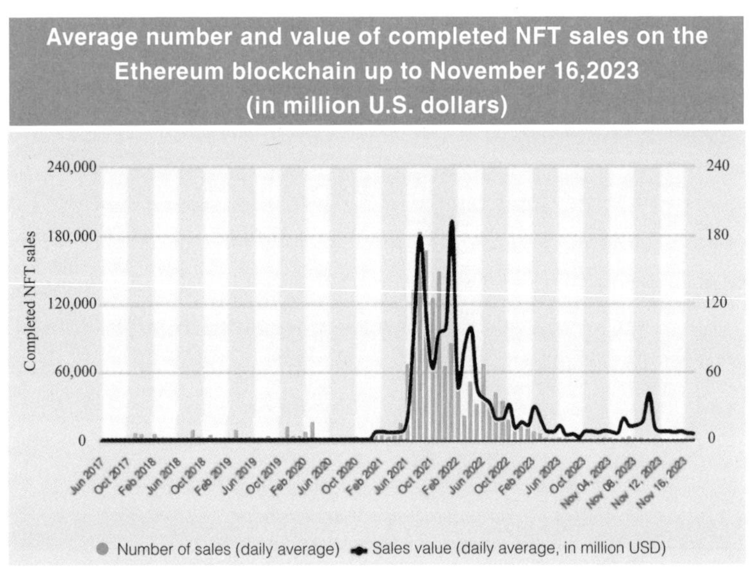

Source: NonFungible

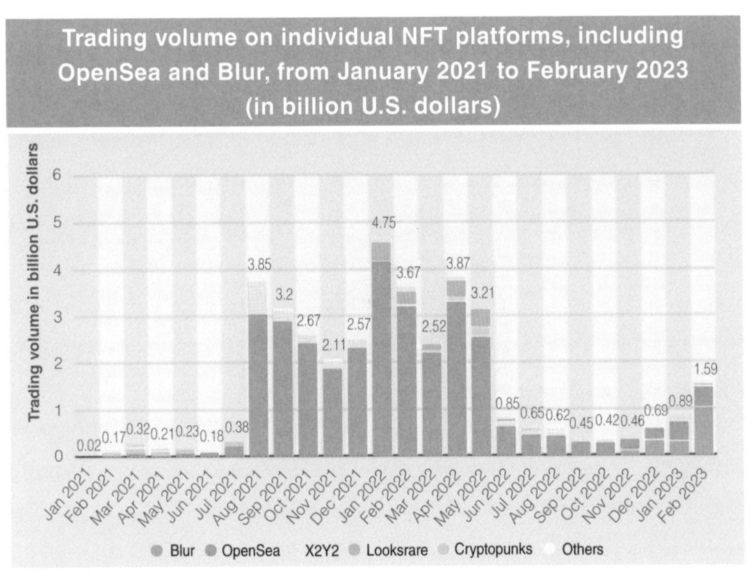

Source: NonFungible

However, despite a significant decrease in secondary market activity and market capitalisation, the minting activity is becoming increasingly active, with more and more users minting NFTs.

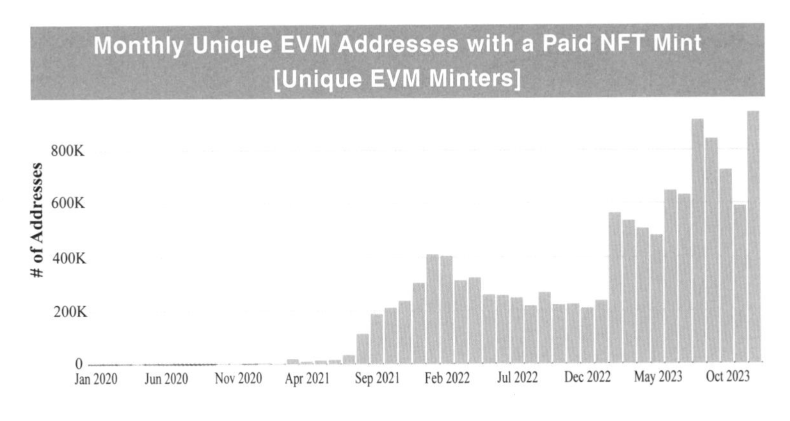

Source: dune.com

6.1 Valuation Matrix of NFT

Since NFTs can represent a wide range of non-fungible items, as mentioned earlier, there are many different categories, making it challenging to have a unified valuation approach. Taking art NFTs as an example, we can temporarily set aside the crypto-native narratives and consider traditional art pricing. However, subjective factors also have a significant influence on the valuation of traditional arts. Also, unlike other financial instruments, most NFTs do not generate stable cash flows, which makes their pricing relatively subjective. Considering these factors, we won't provide a quantitative method in this chapter but will offer a valuation framework for reference.

Utility

Utility emerges as a key parameter in figuring out how to evaluate NFT projects. The applicability or usefulness of an NFT is determined by how it is implemented, be it in physical or virtual spaces. Game assets and tickets are representatives of idols with high practical value. For example, users might need to purchase NFT tickets to attend an art exhibition in Decentraland. And many gaming companies are also considering selling in-game items in the form of NFTs. Players would need to purchase these gaming assets as a form of "admission ticket" in order to participate in the game. The value of NFTs is closely related to their underlying utility, which is easy to understand. For instance, NFT tickets that can unlock access to more exclusive content or experiences tend to have higher prices. Additionally, NFTs with more in-game functionalities are usually priced higher. NFT avatars with visually appealing game skins also tend to command higher prices.

Tangibility

NFTs that have a direct connection to the real world offer a sense of tangibility. By leveraging the secure and immutable nature of blockchain ownership, they provide immediate value that can be physically experienced.

In 2022, Nike and its subsidiary RTFKT Studio launched their first Ethereum-based NFT sneakers, the RTFKT x Nike Dunk Genesis CryptoKicks. According to RTFKT, anyone who owns the Lace Engine NFT has priority access to purchasing Cryptokicks iRL and can enjoy discounted prices. The RTFKT x Nike Dunk Genesis CryptoKicks already include some features such as exclusive physical item casting, whitelist access, and future free casting, making utility an important component of Nike's brand strategy. Through CryptoKicks, they are taking NFTs to a new level by exploring the connection between physical and digital products through exclusive shoes available only to token holders. This was what Nike and RTFKT aimed to achieve with Cryptokicks iRL, which is being touted as "the first real-world Web3 shoe".

Liquidity

Just as with other assets, liquidity premium plays a crucial role in NFT valuation. NFTs with high liquidity carry higher value as well. Factors such as usefulness, previous ownership or sales history, branding, and value appreciation determine the liquidity of NFTs. These components promote customer awareness of NFTs. Traders prefer to put their money in NFT categories with a high trading volume as more liquidity helps them easily exit their positions and reduce the risk of holding. Even when the associated platform is closed, a highly liquid NFT is likely to retain its value as long as there are willing buyers. Low-value and under-recognised NFTs often suffer from poor liquidity, making it difficult for holders to find counterparties when they want to trade.

Example: DeGods

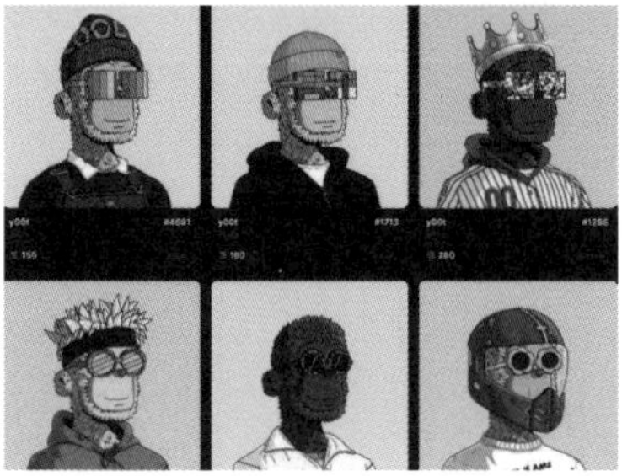

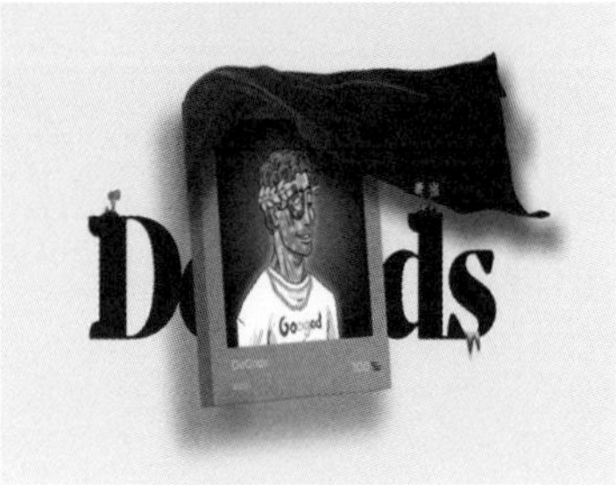

In 2022, DeGods Group announced the transfer of its NFT collections, Degods and y00ts, from Solana to Ethereum and Polygon, respectively. These were previously two of the most popular series in the Solana NFT market. According to data from Magic Eden, in the week leading up to the announcement, sales of DeGods and y00ts accounted for nearly 70% of the total sales volume of Solana NFTs.

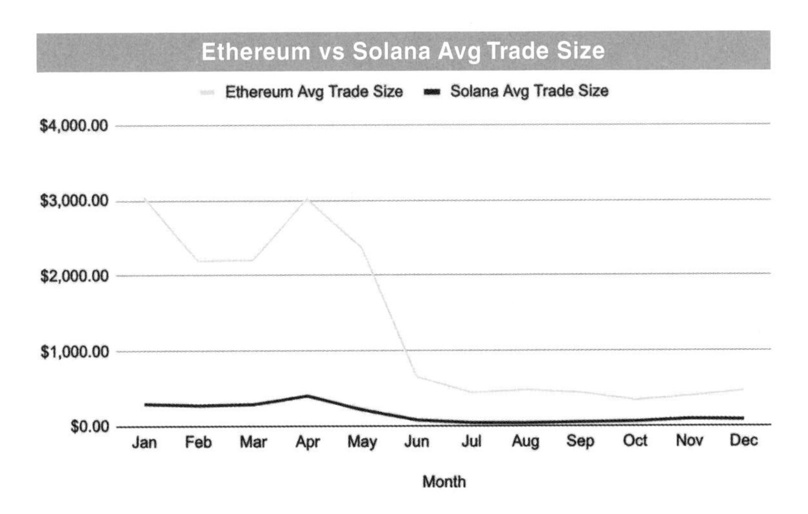

After the announcement of this cross-chain migration, the prices of both NFTs experienced significant changes. Compared to Solana, Ethereum has a significant advantage in terms of DeFi TVL, with trading volume over 100 times higher and an average NFT transaction size five times that of Solana. With the news, sales of DeGods increased, and as of December 26th, the floor price of the collection rose by 12%. Sales of Y00ts remained relatively calm, with the floor price increasing by 5 SOL (approximately $55 at the time).Nansen pointed out that the daily trading volume of DeGods grew by 186%, with the average price rising from 426 SOL to 566 SOL, representing a growth rate of 33%.

In the first half of 2023, DeGods officially migrated from the Solana ecosystem to Ethereum, and their second-generation project, y00ts, also migrated from Solana to Polygon. After migrating both projects, they performed exceptionally well in the NFT winter, with floor prices consistently rising. However, on August 10th of the same year, y00ts would also migrate to Ethereum. Investors seemed to perceive this news as positive. According to OpenSea's data, after the announcement, the floor price of y00ts rose from 1.7 ETH to a peak of 2.1 ETH, with a maximum increase of 23%. As of the time of drafting, it had fallen back to 1.83 ETH, with a recent 8% increase in the floor price and a 352% increase in trading volume over the past 24 hours.

Rarity

Just like in traditional fine art, finding out how rare and difficult it is to possess a given NFT is one technique to determine its worth.

Some NFT collections are more valuable than others, and even within the same collection, not all digital collectibles are created equal. The attributes and traits that an NFT possesses within the collection it represents, among other things, determine its rarity. A one-of-a-kind combination of such characteristics can translate to how difficult it is to obtain a specific NFT, influencing its market value.

According to the basic economic concept of scarcity, an NFT with rare characteristics generally sells for higher prices and is more sought-after. Collectibles with a higher rarity rank may include perks like exclusive memberships and access to limited-edition merchandise.

Generally, the NFT Rarity Score can be used to measure its digital rarity. Several rarity tools to generate NFT Rarity Score have already been published and could be used, such as Rarity Tools. To ensure objectivity and comprehensiveness, rarity ranking scores from multiple sources can be combined.

CryptoPunks is the first batch of blue chip NFTs created by Larva Labs on the Ethereum blockchain. The team drew inspiration from the London Punks scene and cyberpunk movies and novels like Ghost in the Shell to develop a pixel-art character generator.

The total number of CryptoPunks is limited to 10,000, ensuring the scarcity of the collection. Based on their distinct attributes (or lack thereof), some CryptoPunks' portraits may hold more value than others. For example, there are only nine aliens and 24 apes, 44 with top hats, 48 with necklaces, 78 with buck teeth, and 128 with rosy cheeks. 8 CryptoPunks have no distinctive attributes, while one (#8348) possesses seven unique features.

These CryptoPunks with diverse and rare attributes command higher prices compared to those lacking or with more common attributes. For example, CryptoPunk #7804, an alien (9) wearing sunglasses (378), a hat (254), and smoking a pipe (317), was eventually sold for a high price of 4,200 ETH.

Source: Rarity.tools

Interoperability

Interoperability in the context of NFTs refers to the ability of different NFTs and NFT platforms to work together and communicate with each other. This might include the ability to transfer NFTs from one platform to another, or from one game to the other, to use NFTs from one platform as collateral or currency on another platform, or to access information about an NFT on one platform from another platform. Interoperability plays a crucial role in the long-term value of NFTs. However, many NFTs do not prioritise this aspect currently.

The interoperability of NFTs is particularly prominent in blockchain gaming, as it allows individual projects to surpass the limitations of their own ecosystems. For developers, this can expand the appeal of their games, attract a larger player base, encourage greater player engagement, and unlock new business models and revenue streams. For players, it provides more value and utility to their NFTs, encourages exploration of multiple games, enables the implementation of creative gaming strategies across different games, and involves assets from various games.

Celebrity endorsement

Similar to the logic of celebrities or influencers endorsing products in Web2, having the support of celebrities can greatly impact the price and sales volume of an NFT project, especially in its early stages.

Collaborating with individuals or companies with a strong reputation to issue NFTs naturally attracts traffic. For example, the first NFT Formula 1 car sold for a whopping $113,124. Another example is Phanta Bear, which is an NFT collection jointly launched by the art platform Ezek and PHANTACi, a brand under Jay Chou's umbrella. Jay Chou subsequently changed his social media profile picture to Phanta Bear. Under Jay Chou's influence, Phanta Bear was quickly sold out upon its release and topped the NFT sales chart on the same day.

In addition to the perspective of celebrity endorsements and collaborations in the primary market mentioned above, the secondary market is also an aspect to consider. If a celebrity has previously owned an NFT, it is more likely to be sold at a higher price. The ownership history aspect is also one of the references used in the traditional art field for art valuation. Taking BAYC as an example, celebrities like Mark Cuban, Justin Bieber, Stephen Curry, Steve Aoki, Snoop Dogg, Marshall Mathers and Neymar have owned it. However, the sales data shows that most celebrities have not sold their BAYC.

Community Value & Core Culture

Usually, every NFT needs to have a clear cultural value before its release, which can be a specific concept, trend, or social phenomenon. It's important to identify the target audience and establish an emotional connection with the NFT holders, creating a unique symbol and sign for the community. NFTs with specific cultural elements attract focused promotion and operation, continually attracting new members. Well-known NFTs are recognised because they all have distinct themes and target communities. For example, Bored Ape Yacht ClubBAYC) represents existential boredom, CryptoPunks embody crypto fundamentalism and punk spirit, Azuki showcases the culture of ACGN, and World of Women focuses on women's rights. On the contrary, NFTs without a specific cultural core consensus quickly disappear from the market.

In addition to culture, community ecosystem development is also an essential part of a successful NFT. If an NFT can have more use cases, whether it's collaborations with physical products or in virtual worlds such as games, it can increase the exposure of the NFTs and make holders more willing to possess them. Community members also feel a sense of belonging and pride. When it comes to ecosystem expansion, Bored Ape Yacht Club (BAYC) is undoubtedly one of the most outstanding representatives in the entire NFT industry. BAYC members have commercial rights to their owned avatars, and the co-founders have also expressed that creating anything with apes can strengthen the brand further. Yuga Labs, the parent company of BAYC, has built a robust IP ecosystem covering multiple domains, such as fashion, music, food, gaming, and collectibles, with nearly 80 brands, creators, projects, and artists involved.

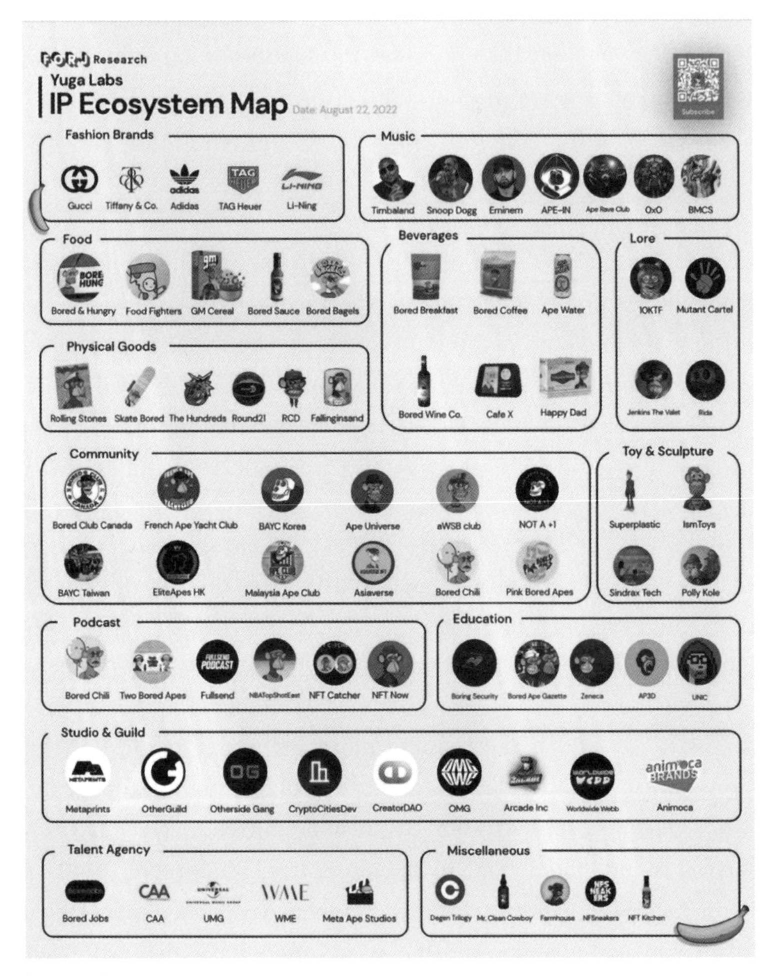

Source: FORJ Research

Social proof

Since NFTs' value is usually subjective and imprecise, social proof wherein people copy the actions of others in choosing how to behave could take place. When people see many users talk about a specific NFT on social media like Instagram or Twitter, they feel like it's a good reason to buy it to incorporate the collective.

In the context of NFTs, social proof plays a significant role in shaping the perception of value and driving pricing trends.

As mentioned above, within NFT communities, there may be a consensus about the value or desirability of certain NFTs or collections. This consensus can shape market behaviour and influence pricing, as collectors look for the opinions of their peers when making buying and selling decisions.

For example, when a collector changes their Twitter profile picture to a BAYC avatar, it immediately attracts followers also with BAYC portfolio pictures. Furthermore, when a BAYC is sold at a high price or held by a celebrity, the entire BAYC collection benefits. The community functions as a collective of shared interests, and all members work together for the success of the NFT. Starting from players changing their Twitter profile pictures to BAYC avatars, their words and actions represent the entire community, serving as a form of social proof.

Security

The security of NFTs is also an important factor for investors in assessing the value of NFTs. A highly secure chain implies a greater level of protection for the ownership and value of NFTs. This increases buyers' trust in NFTs, as they know that their assets are less likely to be stolen or exposed to other security threats. This trust can lead to higher demand and higher prices. In a scenario where multiple blockchain options are available, buyers may be more inclined to choose NFTs on a chain with higher security (Ethereum, for instance). This preference for security can drive up a premium for NFTs on highly secure chains. Conversely, if there are security vulnerabilities or susceptibility to hacking attacks on a particular chain, the prices of NFTs on that chain may decrease. Buyers may approach such NFTs with greater caution, thereby reducing their value.

According to CryptoSlam data, the majority of NFT projects are deployed on Ethereum. While other public blockchain ecosystems also have impressive NFT collections, NFTs on Ethereum indeed enjoy better security and higher liquidity.

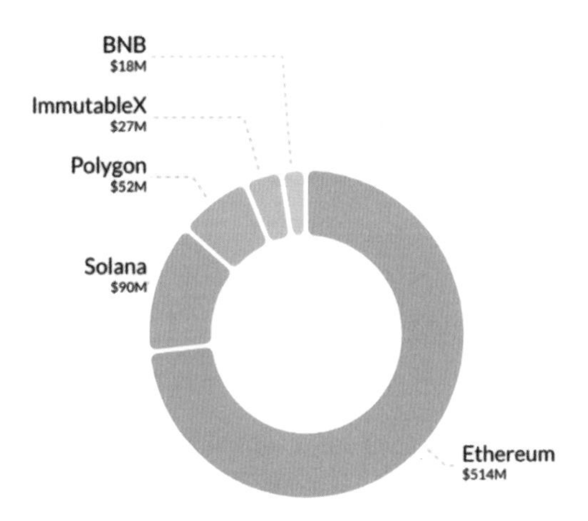

Source: CryptoSlam

BTC NFT

Ordinals and Bitcoin NFTs undoubtedly became hot topics in the crypto field in 2023. Prior to Ordinals, people were more familiar with NFTs on Ethereum. One of the main differences between them is that the metadata of Ordinals' NFTs is stored directly in the Bitcoin blockchain without the ability to introduce off-chain content. Due to the high cost of writing data into the Ethereum Virtual Machine (EVM), storing an image on Ethereum could cost tens of thousands of dollars. In Ethereum, most NFT images themselves are not stored on the blockchain. Instead, only metadata about the images, such as the file's hash value, name, timestamp, and URL link pointing to the file's storage location, are stored on-chain. The actual images may be stored in decentralised storage protocols like IPFS or centralised databases. Additionally, NFTs based on the ERC721 standard on Ethereum are programmable and can be endowed with more interactive functionality. In contrast, the Ordinals protocol only assigns a unified format of transaction data attachment to each UTXO (unspent transaction output), making the data inscription immutable but lacking programmability.

Ordinals

What is Ordinals?The smallest unit of Bitcoin is called a satoshi (sat), and one Bitcoin is composed of 100 million satoshis. Ordinals is a protocol built on the Bitcoin network. Bitcoin developers created an open-source software called ORD, which allows users to assign ordinal numbers to each satoshi using the Ord software. This enables users to observe when satoshis are mined and track them, creating ordinal numbers that make each satoshi unique. Bitcoin is generated through block rewards, and each satoshi is sequentially numbered based on the order of their mining.

Inscription

In addition to assigning ordinal numbers, data can be directly added to these satoshis, and the data can be of any form, including images, videos, and various applications, adding value to the satoshis as collectibles. This process of adding data is called inscription. How is an inscription made possible? The Ordinals protocol can directly write data into the Bitcoin blocks, which is made possible through two upgrades to the Bitcoin network: Segregated Witness (SegWit) in 2017 and Taproot in 2021. The SegWit upgrade effectively increases block capacity by moving signatures and other data that occupy significant storage space to the end of transactions. The Taproot upgrade allows developers to add more complex scripts in the witness area and removes the limitation on the number of operation codes. These two upgrades provide the technical foundation for the Ordinals protocol.

The emergence of the Ordinals protocol has brought new narratives and vitality to the BTC ecosystem. As of December 31, 2023, 52,808,358 inscriptions have been made. The trading volume surged after April, experienced a slight dip in October, and became a point of explosion again in November. Galaxy predicts that by 2025, Bitcoin NFTs could grow into a $4.5 billion market. Blue-chip Ethereum NFT projects like Yuga Labs, Crypto

Punks, and major players like Magic Eden have already joined the Ordinals ecosystem. While opinions within the Bitcoin community vary, some "Bitcoin purists" believe that Bitcoin should adhere as much as possible to its transactional properties as electronic cash and should not add more redundant data. Another part of the community is excited about the additional imagination and space that Ordinals brings to Bitcoin. There is no unified consensus on what Bitcoin should be used for, and no one can define its purpose. In conclusion, the emergence of Ordinals has prompted the community to reconsider the development space of the Bitcoin ecosystem.

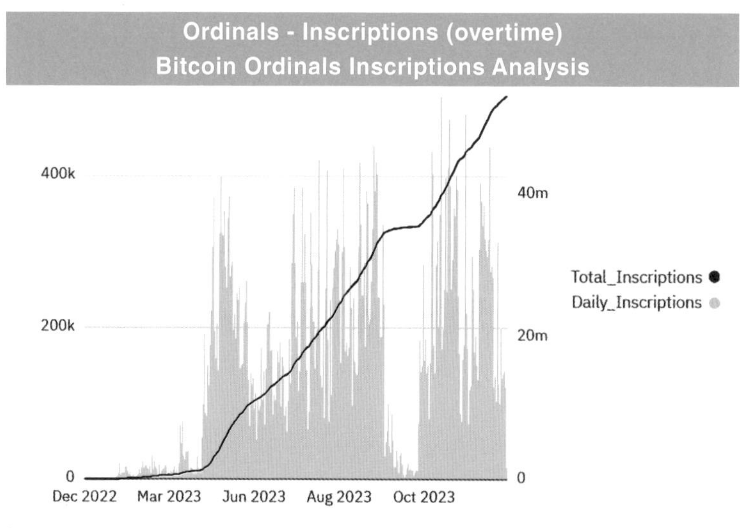

Source: dune.com

NFTs as a hedging tool?

Can NFTs serve as a hedging tool for the entire cryptocurrency market? Regarding this viewpoint, we have conducted some analysis. We performed regression analysis using NFT market cap data and ETH market cap data for the past year (December 28, 2022, to December 28, 2023) and obtained a correlation coefficient of 0.0164, indicating a weak correlation.

Furthermore, by observing the historical data of the NFT market cap and ETH market cap, we can see that the NFT market is less responsive compared to the crypto market. When the market experiences significant fluctuations, NFTs typically do not immediately react. For example, when the ETH market cap decreases, the NFT market cap does not immediately decline and remains relatively stable. However, it is important to note that our analysis is based on data from the past year, and the NFT market is still very young. Therefore, the correlation between NFTs and the entire crypto market is unstable, with significant individual variations. The price movements of individual NFT collections are influenced not only by the overall NFT market and the broader crypto market but also by the narrative associated with each specific NFT.

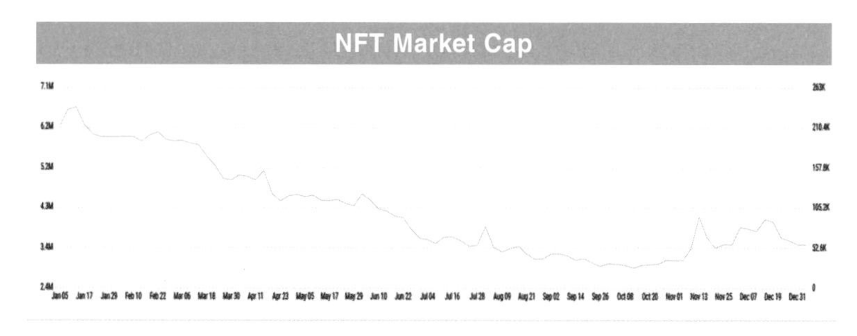

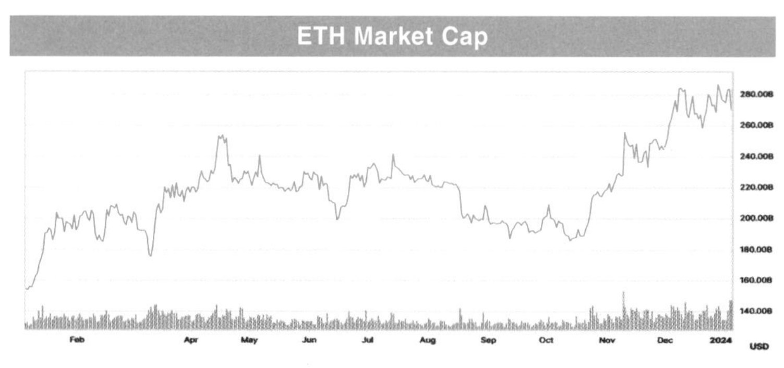

Source: NFTGO & CoinMarketCap

Conclusion

In summary, everything can be an NFT. Although we currently lack a suitable quantitative model to estimate its price, we never deny the value of NFTs. NFTs are not limited to digital art but can support the underlying infrastructure layer of any non-fungible thing. In our real world, non-fungible things often outnumber fungible ones. Leveraging the blockchain's composability, scalability, openness, and censorship resistance, NFTs will serve as an important bridge and tool between the physical and virtual worlds, unlocking limitless potential in the future.

Conclusion

We have been working in the crypto space, and, as any other crypto professional, we often hear questions such as "Why does crypto A or B have value?" "Is Bitcoin backed by anything?" and "Why do crypto prices move?"

With this book, we hope to provide the crypto community, investors, and traditional finance with a valuation framework that allows them to find those answers. Why does crypto have value? What's its fair value? How do we have fundamental metrics for crypto? We hope that these questions can now have more clear answers.

We also hope that the books provide crypto investors with additional valuation tools for their fundamental analysis to level up crypto investments with traditional securities in terms of analytical rigour and understanding.

By equipping investors with these tools, we aim to foster a more informed and rational approach to crypto investment, moving beyond speculation to a more fundamentals-driven perspective.

By presenting these concepts in a structured and accessible manner, we invite not only seasoned investors but also newcomers to the crypto space to deepen their comprehension of what gives these digital assets value. This book is not just for placing on the desks of legendary investors like Warren Buffett but also for anyone who is curious about the underpinnings of crypto asset valuation. In doing so, we strive to bridge the gap between the traditional financial world and the frontier of digital assets, enriching the dialogue and understanding between both spheres.

Although this book attempts to create a comprehensive framework for crypto asset valuation, we recognise that crypto is still a new asset class. We would love to see crypto asset valuation frameworks becoming more and more crystalised in the investment space.

This book can also be seen as a starting point for discussion in the community and financial sphere on how crypto assets should be valued. There are still limitations: crypto is just a little older than ten years, it has limited historical data to backtest our models, and it's a space where innovation happens at the speed of light. However, we are happy to foster the crypto asset valuation in the community.

As the crypto market matures and more data becomes available, these frameworks will undoubtedly be tested and improved upon, leading to more sophisticated and robust valuation techniques. Our hope is that this book contributes to a greater understanding and a more solid foundation for the future of crypto asset valuation.

Acknowledgements: We would definitely like to acknowledge that this book wouldn't be possible if thousands of people hadn't dedicated their lives to developing valuation models in traditional finance. If we attempted to acknowledge all those names, we would not only fail to do it in a fair manner, but it would also take way too many pages from the book.

Finally, we would like to invite the reader to embark on this crypto valuation with us, to engage with the material, challenge the ideas presented, and contribute their own insights to the evolving conversation about crypto asset valuation.